£5.95
8/

SCIENCE AND INDUSTRIALISATION IN THE USSR

STUDIES IN SOVIET HISTORY AND SOCIETY
General Editor: R. W. Davies

The series consists of works by members or associates of the interdisciplinary Centre for Russian and East European Studies of the University of Birmingham, England. Special interests of the Centre include Soviet economic and social history, contemporary Soviet economics and planning, science and technology, sociology and education.

Robert Lewis
SCIENCE AND INDUSTRIALISATION IN THE USSR: INDUSTRIAL RESEARCH AND DEVELOPMENT, 1917–1940

Nicholas Lampert
THE TECHNICAL INTELLIGENTSIA AND THE SOVIET STATE: A STUDY OF SOVIET MANAGERS AND TECHNICIANS, 1928–1935

Further titles in preparation

SCIENCE AND INDUSTRIALISATION IN THE USSR

Industrial Research and Development 1917–1940

Robert Lewis

in association with
Centre for Russian and East European Studies
University of Birmingham

© Robert Lewis 1979

All rights reserved. No part of this publication may be reproduced or transmitted, in any form or by any means, without permission

First published 1979 by
THE MACMILLAN PRESS LTD
*London and Basingstoke
Associated companies in Delhi
Dublin Hong Kong Johannesburg Lagos
Melbourne New York Singapore Tokyo*

*Filmset in Great Britain by
Vantage Photosetting Co. Ltd,
Southampton and London
Printed and bound at
William Clowes & Sons Limited
Beccles and London*

British Library Cataloguing in Publication Data

Lewis, Robert
    Science and industrialisation in the USSR –
    (Studies in Soviet history and society; 1)
    1. Research, Industrial – Russia – History
    I. Title II. University of Birmingham. Centre for Russian and East European Studies   III. Series 607.2'47   T177.R8

ISBN 0-333-23758-7

*This book is sold subject
to the standard conditions
of the Net Book Agreement*

To Margaret, with love

Contents

List of Tables	viii
Preface	ix
List of Abbreviations	xii
1 The Tsarist Legacy	1
2 The Scientific Research Effort of the USSR, 1917–40	6
3 The Industrial Research Effort	20
4 The Central Control of Industrial Research, 1917–30	37
5 Decentralisation and Industrial Research, 1930–40	55
6 The Central Coordination of Industrial Research, 1930–40	67
7 Research Planning	79
8 Science at the Factory	101
9 Industrial Research and Innovation	114
10 Conclusion: Yesterday and Today	143
Appendix 1 The Data for Soviet Expenditure on R&D	151
Appendix 2 A Price Index for Soviet Expenditure in Science	166
Appendix 3 The Data on Soviet R&D Manpower	169
Notes	171
Index	205

List of Tables

2.1 Expenditure on science in the USSR, 1923/24 – 1940   10
2.2 Manpower of R&D organisations in the USSR, 1928 – 1940   13
3.1 Manpower of the USSR's industrial research establishments, 1925 – 1939   21
3.2 Expenditure on industrial research establishments in the USSR, 1923/24 – 1937   23
3.3 Breakdown of manpower employed in the Soviet Union's industrial research institutes (1929, 1931 and 1935) and in the research laboratories of the United States' industrial companies (1927 and 1938).   32
8.1 Factory laboratories in 1933   107
A1.1 Expenditure on 'Science' in the USSR, 1923/24 – 1941   154
A1.2 Expenditure on the industrial research establishments in the USSR, 1923/24 – 1940   157
A1.3 Expenditure on 'Science' in the USSR, 1923/24 – 1941 (summary data and estimates)   162
A1.4 Expenditure on the industrial research establishments in the USSR 1923/24 – 1940 (summary data and estimates)   164
A2.1 A price index for expenditure on Science, 1923/24 – 1940   168

Preface

In recent years there has been increasing discussion, both in the Soviet Union and among western observers, of the Soviet Union's ability to generate new technology from its own resources. This discussion has centred on the USSR's performance in research and development (R&D) and innovation. In particular, there has been considerable debate about the effects of the Soviet system of economic planning and industrial management on technical progress. The organisational structure for R&D has also been held up as partly responsible for the relatively poor performance of the Soviet economy in this respect. In reviewing the historical background to existing planning and administrative systems, many authors refer to the formative years of the Soviet economic system before the Second World War as being similarly crucial in shaping the organisation of R&D and innovation in the Soviet Union today.

This book reviews the development of R&D organisations and the size of the Soviet commitment to science over these years, which saw the construction on the basis of the scientific inheritance from Tsarist Russia of an extensive network of scientific organisations. The growing R&D effort took place against a background of radical economic change as the Soviet economy went from virtual standstill in the years immediately after the revolution to rapid and sustained economic growth as a result of the industrialisation drive which was implemented in the late 1920s. After an assessment of the Tsarist legacy to the Bolshevik government in Chapter 1, the review of the inter-war years begins with an attempt to measure the growth of the Soviet scientific effort as a whole (Chapter 2) and of that part of this effort that was aimed at serving industry (Chapter 3). Chapters 4, 5 and 6 contain a discussion of the administration and coordination of the industrial R&D effort and chart the changes in government policy to the R&D network which was being established. In Chapter 7 the origins and subsequent development of the research planning system are discussed; for scientific research, like other areas of economic life in the

Soviet Union, was seen as an activity which could and should be planned. It was, of course, the idea of planning the development of science for the betterment of society as a whole that was the feature of Soviet science which attracted greatest attention in other countries and which played an important part in fuelling the debate within the scientific community in Britain during the 1930s. We then turn our attention to the factory itself, for a feature of the Soviet R&D system has been the relatively small part played by plants in scientific research, and in Chapter 8 we review the role which was foreseen for the factory in science and the success with which it met the goals set out by the government and the heads of industry. Finally, we must consider the outcome of the policies adopted towards building up an industrial R&D network and of the investment of resources in science. The end of such activity is technical progress and this is achieved by the introduction into industry of the results achieved within the scientific laboratory – innovation. As we shall see, in general, the Soviet Union was not widely introducing new technology which had been developed by its domestic R&D resources, but there were areas in which Soviet-based technical progress did take place. One of these, the aircraft industry, is selected for a closer study as an illustration of the problems surrounding innovation and of the ways in which the Soviet government was in certain circumstances able to solve them.

As is clear from the above remarks this study focuses on Research and Development; however, R&D was not a term in use in the inter-war years. It was a product of the post-war discussions of science and economic growth and the realisation that there was something more than research to technical progress, namely development, the process whereby scientific results are turned into pieces of hardware which can be used as the basis for subsequent large-scale manufacture. In studying the scientific system of the inter-war period, we have attempted to apply the criteria for what constitutes R&D which were developed by the Organisation for Economic Cooperation and Development to try to bring a standardisation to the collection of information on such activities and enable international comparisons to be made of different countries' investments of manpower and funds in such activities. The resulting set of guidelines was entitled the 'Frascati Manual' as result of the adoption of the *Proposed Standard Practice for Surveys of Research and Development* (Paris, 1962) at a conference held in Frascati, Italy, in 1963. As will be seen, in some instances there have been considerable difficulties in fitting Soviet data to the Frascati definitions. A fundamental difference is that in the Soviet Union 'science' (*nauka*) includes the social sciences and humanities, so that expenditure data includes funds spent in these fields and the data on the numbers of scientists (*nauchnye rabotniki*) and total employment in science include workers in these disciplines.

I first started work on the organisation of science in the USSR more than a decade ago at the Centre for Russian and East European Studies of the University of Birmingham and over the subsequent years I have built up a debt to a multitude of people. Firstly there is the large number of former colleagues and friends from Birmingham who have contributed to this book through conversation and by providing references and information. For example, I remember holding long conversations on Soviet science with Pete Kneen, now at the University of Durham, during walks by Devon rivers in spring while Pat and Margaret sadly shook their heads and disappeared into the distance. Secondly there are many scholars in Moscow and Leningrad who have discussed Soviet science with me and helped me to find materials relevant to this study.

I collected material for this book during three visits to the Soviet Union and I would like to thank the SSRC and the Universities of Birmingham and Exeter for financial support on these occasions. However, much of the material consulted has come to me in Birmingham and Exeter through the Inter-Library Loans Service and I am consequently extremely indebted to the Inter-Library Loans Departments of these two universities. I would also like to thank Jenny Brine of the Russian Centre library and the helpful staff at the libraries in Moscow and Leningrad of the Academy of Sciences' Institute of the History of the Natural Sciences and Technology. I am grateful to Elsa and Trudi for typing assistance.

There are, however, two people in particular who must be thanked. The first is Bob Davies, who supervised the thesis on which this book is based, and I hope that it will be seen by him as a fair reward for all the work he has put in. Secondly, my wife Margaret has had to survive my absences in the Soviet Union and in the study, and to add insult to injury has been dragooned into checking drafts and collating pages.

An earlier version of parts of Chapters 2 and 3 appeared in *Science Studies* and parts of Chapters 4 and 6 have similarly appeared in an article in *Minerva*. I gratefully acknowledge the editors' permission to include the relevant material here.

August 1978   R.L.

List of Abbreviations

Cheka	Extraordinary Commission
GEU	Chief Economic Administration (of VSNKh)
Gintsvetmet	State Institute of Non-Ferrous Metallurgy
Giprokhim	State Institute for Projecting Chemical Factories
Gipromez	State Institute for Projecting New Metallurgical Factories
glavk	chief administration (of a Commissariat)
Glavkhim	Chief Administration of the Chemical Industry (of VSNKh and later NKTP)
Glavvtuz	Chief Adminstration of Higher Technical Educational Establishments (of NKTP)
GOELRO	State Commission for the Electrification of Russia
Gosplan	State Planning Commission
GUAP	Chief Administration of the Aviation Industry (of NKTP)
GVF	Civil Air Fleet
KEPS	Committee for the Study of the Natural Productive Forces of Russia
Khimstroi	Chemical Equipment Construction Company
khozraschet	economic accounting
KGB	Committee of State Security
Komsomol	Communist Union of Youth (see also VLKSM)
Narkompros	People's Commissariat of Education
Narkomtorg	People's Commissariat of Trade
NIS	Scientific Research Sector
NKLegProm	People's Commissariat of Light Industry
NKLesProm	People's Commissariat of the Timber Industry
NKMestProm	People's Commissariat of Local Industry
NKPP	People's Commissariat of the Food Industry
NKPS	People's Commissariat of Communications

List of Abbreviations xiii

NKRKI	People's Commissariat of Workers' and Peasants' Inspection
NKSnab	People's Commissariat of Supply
NKTP	People's Commissariat of Heavy Industry
NKVD	People's Commissariat of Internal Affairs
NKZdrav	People's Commissariat of Health
NKZem	People's Commissariat of Agriculture
NTO	Scientific and Technical Department (of VSNKh)
NTS	Scientific and Technical Council (of VSNKh)
NTU	Scientific and Technical Administration (of VSNKh)
ODVF	Society of Friends of the Air Force
OECD	Organisation for Economic Cooperation and Development
OGPU	Unified State Political Administration
Osoaviakhim	Society for the Promotion of Defence, Aviation and Chemistry
PEU	Planning and Economic Administration (of VSNKh)
promfond	fund for financing industry (of VSNKh)
PTEU	Planning, Technical and Economic Administration (of VSNKh)
Rabpros	Union of workers in education and socialist culture
RSFSR	Russian Soviet Federative Socialist Republic
Sovnarkom	Council of People's Commissars (of the USSR, except where stated to be otherwise)
STO	Council of Labour and Defence
SSR	Soviet Socialist Republic
SSSR	Union of Soviet Socialist Republics
TsAGI	Central Aero- and Hydrodynamic Institute
TsIAM	Central Institute of Aero-Engine Construction
TsIK	Central Executive Committee (of the Congress of Soviets)
TsK	Central Committee
TsKB	Central Design Bureau
TsKK	Central Control Commission
TsKK-NKRKI	Joint Committee of the Central Control Commission and the People's Commissariat of Workers' and Peasants' Inspection

TsNIIMash	Central Scientific Research Institute of Engineering
TsUGProm	Central Administration of State Industry (of VSNKh)
VARNITSO	All-Union Association of Workers of Science and Technology to Assist Socialist Construction
VLKSM	All-Union Lenin Communist Union of Youth (Komsomol)
Vsekhimprom	All-Union *Ob"edinenie* of the Chemical Industry
VSNKh	Supreme Council of the National Economy (unless otherwise stated, of the RSFSR, 1917–23; of the USSR, 1923–32)
ZSFSR	Transcaucasian Soviet Federative Socialist Republic

Sources

FNIT	*Front Nauki i Tekhniki*
NR	*Nauchnyi Rabotnik*
Protokol VSNKh SSSR	*Protokol Zasedaniya Prezidiuma Vysshego Soveta Narodnogo Khozyaistva SSSR*
SRIN	*Sotsialisticheskaya Rekonstruktsiya i Nauka*
SU RSFSR	*Sobranie Ukazonenii i Rasporyazhenii Rabochego i Krest'yanskogo Pravitel'stva RSFSR*
SZ SSSR	*Sobranie Zakonov i Rasporyazhenii Rabochego i Krest'yanskogo Pravitel'stva SSSR*
TPG	*Torgovo-Promyshlennaya Gazeta*
Za Ind.	*Za Industrializatsiyu*
ZL	*Zavodskaya Laboratoriya*

1 The Tsarist Legacy

In the twenty-five years before the outbreak of the First World War, there were substantial changes in the organisation of science in the industrially developed countries of western Europe and the United States, with a growing proportion of scientific research now taking place outside the universities. The latter, which earlier in the nineteenth century had taken over from the academies and learned societies as the main institutions in science, were now to see their primacy eroded by a new generation of organisations in the form of research institutes and laboratories.

While the actual pattern of change varied from country to country, these new organisations were, in general, of three types. Firstly, there were growing numbers of industrial research laboratories; these had been established initially in the chemical industry in Germany and they were also to become a prominent feature of electrical-engineering firms. Secondly, some research institutes were established on the basis of private funds to foster a wider range of research than work narrowly directed towards a particular industry; these were, in particular, a growing feature of science in the United States, where they were financed by industrialists such as Mellon and Carnegie. Thirdly, a number of state research establishments were established, such as the Physikalisch-Technische Reichanstalt in Germany and the National Physical Laboratory in the United Kingdom.

At the beginning of the twentieth century the organisation of science in Russia, where substantial industrial development had only recently begun, was still reminiscent of western Europe earlier in the nineteenth century. The preceding years had seen the beginnings of professionalisation in science and the rise to prominence of the universities at the expense of the Imperial Academy of Sciences. Membership of the Academy had ceased to be automatically coveted by university professors.[1] An important reason was its lack of research facilities. Outside the higher education sector, some government-sponsored research was being undertaken, for example under the state's Geological Committee and in agriculture, but much of the work of these organisations was not strictly classifiable as R&D. There was also some private funding of science. The newly established Institute of Experimental Medicine to which Pavlov, the famous physiologist, moved in 1890 had been founded by Prince A.P. Ol'denburgskii.[2]

In its use of science Russian industry lagged far behind industry in western Europe and the United States. A growing number of enterprises did have laboratories, but the factory scientists employed in them were largely engaged in the scientific servicing of production. Only in a few cases, such as at the Putilov plant in St Petersburg, was research undertaken in addition to day-to-day testing and analysis.[3]

During the early years of the twentieth century the development of scientific organisations in Russia continued to lag behind the more industrialised countries. At the outbreak of the First World War there were still few R&D organisations in the Empire. Russia's scientists, however, were not unaware of events abroad. Those who observed the growing role of specialised research institutes in other countries and the part played by their governments in establishing research facilities pressed for a start to be made on the creation of a Russian research network.[4] They found the government unresponsive to requests for financial help in establishing research institutes. In 1913 the state budget allocated about 9 million out of a total budget of 3400 million rubles to scientific establishments and for scientific work.[5] Just under half these funds were put at the disposal of the Ministry of Education and comprised less than three per cent of its total resources. A large part of the remainder was spent by the government departments responsible for agriculture. Over 500,000 rubles of the Ministry of Education's expenditure was on museums and archives, which by their nature were not involved in research to any extent, and a further million rubles funded the Imperial Academy of Sciences, which remained more a scientists' club than a research organisation and was largely active in the humanities. Indeed, the ministry's attitude towards science was attacked in the following year in a critique published by the Progressive Faction – a moderate liberal group – of the State Duma on its budgetary proposals; this noted that very little money was to be allotted for new scientific needs and that neither in 1914 nor in the preceding years had any new scientific establishments been founded by the ministry.[6]

Private backing for research was also hard to acquire. Russian industry did little to match the continuing growth of factory research laboratories in other countries. An important factor in explaining this lag was undoubtedly the disproportionate role played by foreign capital in the development of industry;[7] as a result the technology of Russian industry depended on foreign firms and R&D was imported in its final form as machinery and equipment. However, by the First World War there was an increasing domestic involvement in industries which had previously been dominated by foreigners[8] and there appears to have been little willingness on the part of domestic entrepreneurs to foster science at their industrial enterprises nor did Russian industrialists – in contrast, for example, to their German counterparts – finance any substantial amount of research at higher educational establishments. A handful of small private research organisations were opened in the period after 1905; these were mainly concerned with the study of the country's natural resources.[9] Clearly a wide gap existed between Russian industry and Russian science and there were several examples of Russian inventors being forced to go abroad to find support for their ideas as financial backing was not forthcoming in Russia.[10]

This lack of a growth of independent research organisations was to heighten the role of the higher education establishments as the major source of scientific activity. Indeed there was a large expansion in higher education from the end of the nineteenth century. However, as a consequence of its view of the universities as centres of dissent and rebellion, the government particularly encouraged the development of specialised professional educational establishments, and these grew much faster than the universities. On the other hand, although there was some expansion in the numbers of laboratories at the specialised technical institutes which were created after 1890, scientific research was mainly concentrated in the universities. The stormy relations between the state and the universities did not provide an atmosphere conducive to any commitment on the part of the Ministry of Education to develop scientific research. V. I. Vernadskii appears to have expressed a view held by many scientists at the turn of the century, when he wrote 'that the research activities of the university professors were carried out not according to but in spite of the will of the government'.[11] Scientific facilities in the universities did improve in the decade before the First World War, but this was part of a plan to keep students so involved in their studies that they could not take an active part in politics.[12]

With the state and private industry both failing to provide the level of systematic support desired by Russian scientists, the latter were left to seek out whatever support they could find. Their main fund-raising agencies were the scientific societies which had sprung up in the second half of the nineteenth century. The societies did, in fact, receive some government subsidies and in some cases help from industry, but they were primarily supported by private donations, membership fees and income from their publications.[13] However, the vast majority were established as and mainly remained forums for information exchange and discussion and only occasionally aided research.[14] Among the exceptions were societies set up with the expressed aim of fostering research. The first such society was formed in 1909 and financed by a private donor, the industrialist Kh. S. Ledentsov, with the title of the Kh. S. Ledentsov Society for Advancing the Experimental Sciences and their Practical Application. Some of its funds went to finance the equipping of a laboratory for Pavlov and for N. E. Zhukovskii's aeronautical research at the Moscow Higher Technical College. Two years later a similar organisation was established to collect funds through subscriptions to finance the creation of new research institutes – the Society for a Moscow Scientific Institute.[15]

The total size of the available public and private finance was far from great enough to provide the necessary amount of research facilities for the country's scientists. Leading scientists often worked more or less by themselves in small personal laboratories, not uncommonly in their own homes.[16]

The outbreak of the First World War enforced a change in the government's attitude to science. One of its immediate effects was that Russia was cut off from Germany, which had supplied much of her technology and many vital raw materials. The realisation of the greatness of this dependence led to almost panic measures in Russia, with the piecemeal creation of R&D facilities. Responsibility for the development of industries linked to defence came under military control and the War Department created R&D organisations to deal with problems of interest to them. Thus, there was a rapid development in the work of the War Department's Central Scientific and Technical Laboratory, which was in the process of being organised when war broke out, and an experimental factory for chemicals was set up in Petrograd in 1916 under the War Chemical Committee.

The scientists who had been agitating for the development of the country's research organisations now received a more sympathetic hearing. In 1915 those Academicians who had formulated plans for the organised study of Russia's natural resources and the formation of research institutes in the necessary specialities saw some of their ideas come to fruition with the creation in the Academy of Sciences of the Commission for the Study of the Natural Productive Forces of Russia

(KEPS) under the chairmanship of Vernadskii. This was the first organisation in Russia to draw up overall research plans and to coordinate the work of the various types of research bodies. On the other hand, the funds for its work were very limited and the scientists were unable to realise their full plans.[17] Vernadskii continued to press for a start to be made on the formation of a state research network.[18] Nevertheless, several of the subcommittees and divisions formed within KEPS to deal with specific topics did valuable work and were to form the basis for later research institutes.

Private industry, however, would appear to have shown little sign of a changed attitude to science. In 1917 A. E. Chichibabin, a noted chemist, spoke of the almost complete failure of industry to realise the need for a close connection between science and industry and the absence of any concrete links between them.[19]

It seems clear, therefore, that in pre-revolutionary Russia the organisation of R&D lagged behind western Europe and the United States. In spite of efforts made during the First World War there were relatively few research institutes or equivalent bodies, and little in the way of R&D facilities in industry itself. On the eve of the revolution of October 1917, somewhat over four thousand scientists were occupied on research.[20] Notwithstanding the handicaps facing organised research, the work of such scientists as Mendeleev, Lebedev, Pavlov and their pupils and followers put Russian science to the fore in many fields.[21] As Vucinich writes, 'the scientific community in Tsarist Russia, though labouring under many handicaps, had laid a solid foundation for the notable achievements of Soviet science'.[22]

2 The Scientific Research Effort of the USSR, 1917–40

The majority of scientists while not, perhaps, actively hostile to the new Bolshevik government, undoubtedly distrusted it. The Bolsheviks, for their part, were equally distrustful of the scientists, part of the 'bourgeois intelligentsia'. They were, on the other hand, enthusiastically committed to the development of science and technology to which they attributed an important role in building the new society. Lenin realised that the scientists and technologists had a vital part to play in the future development of the country; speaking in Moscow in April 1918, he said: 'we need their [the bourgeois specialists'] knowledge, their skills, their labour'.[1] His close interest in science was reflected in his oft-reprinted 'Outlines for a plan for scientific and technical work' of the same month.[2] The Party Programme approved at the Eighth Party Congress (March 1919), which spoke of striving for the further development of the country's scientific resources and for the establishment of the most fruitful conditions for scientific work, simultaneously mentioned the need to make the greatest use of the scientific and technical specialists 'in spite of the fact that they, in the majority of cases, are inevitably impregnated with bourgeois attitudes and habits'.[3] In December, F. E. Dzerzhinskii, head of the Cheka – the secret police – published an order which stated that, although the majority of scientists were of a bourgeois outlook, repressive measures should only be taken against them when it had been precisely established that they had been working to overthrow the government.[4] Special provisions were in fact made to improve the living conditions of scientists and ensure that they had sufficient living space to enable them to work.[5] The plan for the comprehensive electrification of the country drawn up by GOELRO, the State Commission for Electrification, which had Lenin's enthusiastic support, provided further evidence of the government's conviction of the importance of science and technology.⁶

The government gave an encouraging response to the plans of scientists for new research establishments which had been previously frustrated and new research institutes were founded on the basis of research groups already in existence in the higher educational establishments and attached to the scientific commissions which had been set up during the war. The Central Aero and Hydrodynamic Institute (TsAGI), for example, was established in 1918 on the basis of N. E. Zhukovskii's aeronautical research group at the Moscow Higher Technical College.⁷ In 1918 and 1919 a total of 33 research institutes were founded; ⁸ these included the institutes which were later to achieve renown as A. F. Ioffe's Leningrad Physical Technical Institute and the Karpov Chemical Institute. These new institutes, the existing state research establishments and the few private research establishments which had been nationalised comprised the initial core of the Soviet research system. However, the issuing of decrees setting up scientific establishments or nationalising existing organisations did not ensure the provision of the funds needed to carry out the plans of the scientists. The years immediately after 1917, marked by a civil war and massive economic problems culminating in hyperinflation, were a period of extreme hardship. While the successor of the Tsarist ministry for education, the new People's Commissariat of Education, Narkompros, spent 37.5 million rubles on science in 1918 and 168 million in 1919,⁹ and these sums appear enormous when compared with the 1913 expenditure of its Tsarist counterpart, prices had risen dramatically in the intervening years,¹⁰ and these funds represented a smaller proportion of the expenditure of Narkompros (just under 1 per cent in 1919¹¹) than the share of its budget which the Ministry of Education had allocated to the scientific establishments in 1913. Further, Narkompros received a relatively small share of total government expenditure.¹² This had obvious consequences for its support of scientific activity in spite of the commissariat's sympathy with the needs of scientists.¹³ Funds appear to have been available to a somewhat greater degree from VSNKh, the Supreme Council of the National Economy, which was responsible for industry.¹⁴ Scientific establishments suffered from severe shortages of resources. In 1920 and 1921 the Academy of Sciences wrote several times to Sovnarkom, the Council of People's Commissars of the RSFSR, on the critical state of Russian science and scientists¹⁵ – between 1918 and 1920 twelve Academicians died.¹⁶

From 1922 the Soviet economy began to emerge from the economic abyss. Industry began to recover under the New Economic Policy and the currency reform of 1924 was accompanied by a stabilisation of the budget. By the mid-1920s the various branches of the economy were approaching pre-war production levels. There was a simultaneous increase in central planning and growth in capital investment which culminated in the First Five Year Plan.[17]

A corollary of the capital investment was technical development.[18] The correspondingly important role of science and technology was widely recognised. At the Third Congress of Soviets of the USSR in May 1925, Rykov, the chairman of the Sovnarkom for the USSR which had been established on its formation in 1923, referred in his report on the work of the government to the role of science and technology in the future development of the country in the following terms:

> ... in order quickly to advance the economy to overcome poverty, ignorance and backwardness in the city and the countryside, we must apply science and western European technology in all branches of the economy.[19]

Dzerzhinskii, who became head of VSNKh in 1924, contrasted the situation with the period of recovery when the Soviet Union had been 'overloaded with immediate problems, with the creation of the minimal conditions for development' and had not been able to pay attention to the 'basic and vital' problem of science and technology.[20] He personally took a very close interest in the research institutes under VSNKh and played a key role in discussions on the development of its research network and the best form of organisation to meet the demands of this new economic situation.[21]

The utilisation of the achievements of contemporary science was further seen as being of great importance in improving industrial efficiency. The Central Committee of the Communist Party in March 1927 approved a resolution on the 'rationalisation of production' which spoke of the importance of the country's scientific and technical establishments in raising the technical level of industry.[22]

Indeed one belief which the three leading protagonists of the industrialisation debate – Stalin, Trotsky and Bukharin – held in common was the power of modern science and technology. Stalin's famous use of the phrase 'to catch up and overtake the capitalist countries in a technological and an economic respect'[23] as a formulation of the aim of the Soviet Union suggests a need to provide resources not only for the improvement and further development of western technology but also for the development of domestic technology. Trotsky and Bukharin were more explicit as regards the role of science.[24] The former in a speech made in 1926 stated that the Soviet Union 'had to catch up with the most advanced countries in the field of the latest scientific and technical achievements'.[25] In a 1927 article entitled 'Science and the USSR' Bukharin stressed the socialist planned economy's need for a huge expansion in scientific work.[26]

However, while there was widespread reference to the importance of science, there was no specific overall policy for the development of science, no central government body which coordinated the R&D efforts of the various commissariats and departments. The concrete implementation of the general emphasis on developing scientific research rested with the individual commissariats and departments. The major role *vis-à-vis* science at the highest level of government was played by Sovnarkom USSR. It was directly responsible for the Academy of Sciences, had ultimate authority over the organisation of research institutes by all-union commissariats[27] and reviewed the commissariats' budgetary proposals for scientific research.[28] In 1927, for example, it raised the projected expenditure on VSNKh's research establishments by 20 per cent and the funds that were to go to its own research establishments by one third.[29] The work of Gosplan, the State Planning Commission, did not include the preparation of plans for science until after a Sovnarkom resolution of 8 June 1927, which expanded its role to include the field of social and cultural development.[30]

Over the years between 1924 and 1928 there was a substantial growth in the research effort. Direct budgetary expenditure on science grew from under 20 million rubles[31] to over 80 million and there was an increasing amount of indirect finance for research establishments. Total expenditure probably grew about five times in real terms (see Table 2.1). No data are available on the increase in manpower in these years. However, both expenditure and manpower in the research establishments under VSNKh increased by four times.[32] This suggests that there was probably at least a general fourfold expansion in the research effort of the Soviet Union – albeit from a small initial size.

The discussion of the reports on the projected five year plan for the development of the national economy which took place at the Fifteenth Party Congress in December 1927 was the first occasion at a party congress or conference when more than a passing mention was made of science.[33] Several speakers, including Chubar' (the head of

TABLE 2.1 Expenditure on science in the USSR, 1923/24–1940 (million rubles)

	Budgetary expenditure (current prices)	Non-budgetary expenditure (current prices)	Total expenditure (current prices)	Total expenditure Index at 1926/27 prices (1927/28=100)
1923/24	(16)	1	(17)	(20)
1924/25	(25)	1	(26)	(28)
1925/26	40	n.a.	n.a.	n.a.
1926/27	54	n.a.	n.a.	n.a.
1927/28	81	36	117	100
1928/29	103	62	165	130
1929/30	216	90	306	219
1931*	257	238	495	306
1932	308	(342)	(650)	(333)
1933	344	370	714	339
1934	434	449	882	363
1935	678	415	1093	367
1936	798	(400)	(1200)	(335)
1937	853	(500)	(1350)	(346)
1938	(900)	n.a.	n.a.	n.a.
1939	(1000)	n.a.	n.a.	n.a.
1940	1135	n.a.	3000	588

n.a.=not available.
Figures enclosed by ( ) involve substantial risk of error.
* In 1930 a special budget was drawn up for the period October to December to change the budgetary year to coincide with the calendar year.
SOURCE Appendix 1, Table A1.3.

the government of the Ukraine)[34] and Voroshilov[35] referred to the successful work of the various research institutes. On the other hand in the materials presented to the congress on the plan there were no details of the planned future development of the country's research network. M. N. Pokrovskii, the historian, who was a deputy commissar of education, found it strange that in spite of the numerous statements on the role of science in socialist construction, he had been unable to find a section dealing with scientific development; yet, he pointed out, the plan was so detailed as to provide data on 'the import of merino ewes and the manufacture of refrigerators'.[36] The absence of information on the development of science can be contrasted with the role attributed to it in the resolution which the congress passed on the five year plan. It picked out seven fields as being vital to the success of the plan. The greatest importance was attached to 'the most energetic and strenuous work on rationalisation (of industry, of the trade and state apparatuses etc.)'. The resolution continued:

This rationalisation cannot be done without raising the role of *science* and *scientific technology* (*nauchnaya tekhnika*) [italics in source – R.L.]. The broad development of the network of industrial trial research institutes and factory laboratories, the bringing of academic scientific work into contact with industry and agriculture, the widest utilisation of western European and American scientific and scientific-industrial experience, the detailed study of all the very latest discoveries and inventions . . . must be adopted as the next tasks.[37]

In connection with the work on rationalisation the congress further instructed the Central Control Commission 'to check the correctness of the utilisation of our scientific and technical manpower'.[38] This, in fact, resulted in a detailed study of the industrial research establishments, which formed the basis of a Sovnarkom decree of August 1928 on the organisation of research for the needs of industry.[39]

It was evident from remarks made by Pokrovskii at the Sixteenth Party Conference in April 1929 that the growing emphasis on science was already producing concrete if not entirely beneficial results. He pointed out that any self-respecting commissariat now wanted to have 'its science and its research establishments'; but the result was, as he continued, that:

if the commissariat is rich it has many of them and builds premises for them which perhaps will rival the Neskuchnyi Palace.[40] If the commissariat is poor it builds a boxroom (*chulan*), puts two scientists in it and still hangs out the sign 'Research Institute'.[41]

The First Five Year Plan was approved by the Sixteenth Party Conference and made law by the Fifth Congress of Soviets of the USSR, which met a month later. No global figure for the planned development of research appeared in the published material on the plan,[42] but at the Fifth Congress of Soviets a figure of two milliard rubles for the projected expenditure on research was given by Krzhizhanovskii, the head of Gosplan, in his report on the plan. When summing up, he spoke of scientific research as being 'rather a bottleneck'.[43] In fact the total expenditure on science in the First Five Year Plan – it was declared completed after four and one quarter years at the end of 1932 – did not reach Krzhizhanovskii's figure. Nevertheless a further rapid expansion in the research effort occurred. The total expenditure in the period was just over 1.6 milliard rubles, and by 1932 annual expenditure at about 650 million rubles was three and a half times the 1927/28 expenditure after allowing for salary and price increases (see Table 2.1); over the same period capital investment in the Soviet Union had a little more than doubled.[44] The number of scientists employed in research establishments had grown by two and a half times between the spring of 1929 and the beginning of 1932 and was still increasing (see Table 2.2), while total manpower was perhaps growing faster still. The number of engineering and technical personnel in Soviet industry grew threefold over these years.[45]

By the end of 1932 signs of strain were appearing in the economy and in 1933 the amount of capital investment actually fell.[46] The scientific effort had also overstretched itself. The continuing desire of individual departments to have their own research facilities, of which Pokrovskii had spoken four years earlier, had resulted in a continuous expansion in the numbers of research institutes, from around 400 at the beginning of 1929 to 940 at the beginning of 1933;[47] there was simultaneous pressure on existing institutes to found numerous branches. The result was a proliferation of ill-equipped and poorly staffed research establishments. Towards the end of 1932 the state control organ, the Central Control Commission, reviewed the research network.[48] There followed a drastic reorganisation at the beginning of 1933. Many establishments were liquidated or amalgamated with others; the number of research institutes was cut from 940 to 840, their branches from 303 to 188.[49] Below-standard personnel were fired or down-graded and the number of scientists dropped by 10 per cent (see Table 2.2).

The Second Five Year Plan envisaged that after cuts in manpower in 1933 the number of scientists employed in research institutes – and also the number of institutes – would start to grow again from 1934. This growth was to be at a much slower rate than during the first plan, so that in 1936 the number of research institutes would still only be 626 and they would employ only 46,000 scientists.[50] In fact, while the level of cuts in the initial years of the plan were not as severe as forecast, the network of research institutes was reduced throughout the second plan and for most of the rest of the period under review. In 1939 there were 694 institutes (with 63 branches),[51] which employed 34,600 scientists. These scientists were being supported by an increasing army of technical and auxiliary staff – a growth of 40 per cent between 1935 and 1939 (see Table 2.2). However, the research institutes represented a declining proportion of the R&D network. The total number of all research establishments at the beginning of 1941 showed a considerable in-

TABLE 2.2 Manpower of R&D organisations in the USSR, 1928–40[a] (beginning of year except where stated, thousands)

	1928	1929	1931	1932	1933 (i)	1933 (ii)	1934	1935	1936	1937	1939	1940
Research institutes and their branches	—	—	57·4[de4]	—	—	—	95·8[d11]	90·9[d13]	—	—	108·5[d17]	—
of which: scientists	—	—	31·6[5]	42·2[7]	53·0[9]	47·9[7]	41·3[5]	38·2[5]	39·1[15]	37·6[15]	34·6[15]	—
Research establishments	—	40·2[cd2]	78·7[de6]	—	—	—	—	153·5[c14]	175·0[c16]	—	—	—
of which: scientists	—	18·2[cd3]	34·7[6]	47·6[8]	—	51·1[10]	48·3[12]	45·9[5]	—	—	—	—
Project and design organisations	—	—	—	—	—	—	—	71·3[c14]	96·2[c16]	—	—	—
All R&D organisations	64·0[b1]	—	—	145·0[b1]	—	—	—	—	—	234·0[b1]	—	268·0[b1]

Notes

[a] The data in this table are mainly for the number of posts; they exaggerate the actual manpower in the first half of the 1930s due to the existence of 'multiple-post-holding' (see Appendix 3).
[b] Annual average employment.
[c] 1 April.
[d] Survey data, some organisations not included.
[e] 1 March.

See p. 14 for sources.

14 Science and Industrialisation in the USSR

SOURCES TO TABLE 2.2

1. *Trud v SSSR* (Moscow, 1968) pp. 24–5.
2. *Nauchnye Kadry i Nauchno-Issledovatel'skie Uchrezhdeniya SSSR* (Moscow, 1930) p.72; the survey was considered to have covered 90 per cent of all research establishments.
3. Ibid., p.17.
4. *Narodnoe Khozyaistvo SSSR* (Moscow-Leningrad, 1932) p. 546; this survey may only cover research institutes; the number of scientists is given as 22,500.
5. *Sotsialisticheskoe Stroitel'stvo SSSR* (Moscow, 1936) p. 589.
6. I.S. Samokhvalov, 'Chislennost' i Sostav Nauchnykh Rabotnikov SSSR', *SRIN*, no. 1 (1934) p. 128; the figure for scientists is stated to be a Gosplan figure; the survey figure for the number of scientists in research establishments was 31,500.
7. *Sotsialisticheskoe Stroitel'stvo SSSR* (Moscow, 1935) p. 624.
8. *FNIT*, no. 4–5 (1932) p. 129.
9. *Sotsialisticheskoe Stroitel'stvo SSSR* (Moscow, 1934) p. 418.
10. Samokhvalov, *SRIN*, no. 1 (1934) p. 131.
11. *Kul'turnoe Stroitel'stvo SSSR v Tsifrakh: ot VI k VII S"ezdu Sovetov (1930–1934gg)* (Moscow, 1936) p. 148; a survey of 703 (out of 860) institutes, 118 (out of 190) branches; the source gives a figure of 34,800 for scientists.
12. Addition of data for research institutes (and their branches) and other research establishments from *Sotsialisticheskoe Stroitel'stvo* (1936) p. 589 and *Kult'urnoe Stroitel'stvo SSSR . . . . (1930–1934)* p. 143.
13. *Kul'turnoe Stroitel'stvo SSSR. 1935* (Moscow, 1936) p. 226; data for 847 (out of 933) institutes and branches; the survey figure for scientists was 34,400.
14. *Trud v SSSR* (Moscow, 1936) pp. 26–31.
15. *Kul'turnoe Stroitel'stvo SSSR* (Moscow, 1940) p. 230.
16. *Chislennost' i Zarabotnaya Plata Rabochikh i Sluzhashchikh v SSSR* (Moscow, 1936) pp. 8–13; the large increase in employment in project and design organisations over the previous year is stated to be the result of the inclusion in 1936 of project organisations serving the construction industry which had not previously been included under this head.
17. *Kul'turnoe Stroitel'stvo SSSR* (1940) p. 238; data from a survey of 684 (out of 694) institutes and 61 (out of 63) branches; the survey's figure for scientists and 'engineers and specialists doing scientific work' was 34,200.

crease over 1934[52] and the data published in recent Soviet statistical handbooks show that the total employment in all the various types of R&D organisation nearly doubled in the years between 1932 and 1940 (see Table 2.2).

Further, we have as yet not considered the role of the higher educational establishments in research; in other countries these were very important research centres. In the Soviet Union they were very much overshadowed by the development of the research institutes. Indeed, at the time of the great expansion in the numbers of research institutes during the First Five Year Plan many of the better research centres and laboratories in educational establishments would appear to have been utilised to form the core of a new institute.[53] These years were also the period of greatest pressure on the teaching role of higher education, with a vast growth in students and a great pressure to turn out specialists for the industrialisation drive. This concentration on teaching at the expense of research was to some extent officially sanctioned by a separation of research from teaching which was envisaged in the 1932 statute for the universities of the RSFSR. This statute envisaged that university research would take place in research institutes which would be part of the university but under the direction of the scientific sector of the commissariat itself.[54] However, the concentration on teaching to the exclusion of research was being attacked from the early 1930s by a growing lobby of scientists who sought to point out the catastrophic effect of a complete lack of research in the educational establishments.[55] Nevertheless possibly only about four million rubles of the 1935 state budget was spent on research at the universities and technical institutes.[56] They were also receiving unknown but probably quite small funds from contracts made with various organisations for particular pieces of work.[57] The years after 1935 saw an attempt to reverse the trend of the previous years with an increasing stress on universities and technical institutes as bodies which not only trained specialists but also undertook research. In the second half of the 1930s some establishments acquired the resources of research institutes which were closed down.[58] A reorganisation of the pay system for teachers in higher education at the end of 1937 was particularly aimed at increasing the amount of research undertaken by them and it envisaged that staff should spend half of their time teaching, half on research.[59] However, in practice the universities and the technical institutes remained very much teaching bodies and the worry about the lack of research continued with simultaneous further discussion of possible institutional measures to increase their research activity.[60] Nevertheless there were at the end of the 1930s 60,000 scientists in the higher education sector,[61] so that even if only one-eighth of their time had been spent on research their total input would have been equivalent to nearly one-quarter of the full time research scientists in the institutes.

An immediate effect of the reorganisation in the research network at the beginning of 1933 was that, while in the years from 1929 to the end of 1932 the growth of manpower had clearly outstripped the growth in funds, expenditure per scientist was now rising. This reflected the urgent need to improve facilities and equipment for research. This was emphasised in the Second Five Year Plan which foresaw that capital investment in science would be three times that of the years of the first plan.[62] This was part of a growing stress on the need to improve the quality of research in order to get the fullest return from the R&D network.[63] However, the doubling in the annual rate of funding which the plan envisaged was far from achieved; expenditure in real terms in 1937 was possibly even less than in 1932 (see Table 2.1).

16 Science and Industrialisation in the USSR

There would appear to have been substantial growth in expenditure in the later years of the 1930s, but this may be exaggerated by differences in the range of expenditures included in the published data.[64] Even so it did not grow as fast as national income, which according to the official Soviet series almost doubled in the second half of the 1930s.[65] Further, on the basis of the data published in post-war Soviet statistical handbooks it would certainly appear that there was little increase in real expenditure per head over the whole period between 1932 and 1940 (see Tables 2.1 and 2.2).

THE RELATIVE SIZE OF SOVIET EXPENDITURE ON R&D

The most usual measure of the effort a country is putting into R&D is the proportion of its resources which it is prepared to devote to it – the percentage of its National Income spent on R&D activities. To attempt this calculation for the Soviet Union, we have first to separate expenditure on R&D from expenditure on 'science'. Besides including expenditure on the Social Sciences and Humanities the budgetary category 'science' includes expenditure on non-R&D organisations such as museums, libraries and archives, and on the training of personnel at research establishments. Furthermore, it is necessary to add expenditure on R&D which does not appear under this category in the budget, such as that on research work in the higher educational establishments.[66] Using detailed material which is available for 1935, it would appear that the expenditure through the budget on R&D organisations and on R&D undertaken elsewhere was 500 million rubles, as compared with the 600 million listed as spent on 'science'.[67] We also know that 415 million rubles on additional non-budgetary funds went to research establishments. Assuming that this money went only to those organisations engaged in R&D, total expenditure was 915 million rubles.

What proportion of Soviet National Income in 1935 did this represent? Official Soviet figures in current 1935 prices do not exist; they are only available in 1926/27 prices.[68] The estimated 1935 expenditure on R&D of 915 million rubles thus has to be deflated for changes in wage and salary levels and in that year our price index for expenditure on science stood at 273 (1926/27 = 100).[69] After adjusting the expenditure figure to account for these rising costs, we find that it represents 0.6 per cent of the official Soviet figure for National Income in 1935. However, three additional factors must be taken into consideration; of these, the first suggests that this figure is too high, the others that it is too low.

Firstly, for the purpose of this calculation the assumption has been made that all expenditure on R&D organisations was spent on R&D. As will be shown in the following chapter, a substantial proportion of their work was not in practice R&D. Secondly, our figure of 915 million rubles does not include expenditure on R&D through the defence budget, nor funds for R&D in factories and in independent design and development organisations under the industrial commissariats.[70] Finally, western economists consider that Soviet National Income statistics in 1926/27 prices overestimate the Soviet Union's rate of growth; their main criticism is that machinery and equipment produced for the first time after this base year tends to be overvalued. Jasny, one of the most critical, used his own 1926/27 prices. While his calculated figure for 1928 agreed with the official figure, for 1937 his estimate was only 55 per cent of the Soviet estimate.[71] Since western price indices have been used to deflate the proportion of the funds of the research establishments spent on equipment (and the major part of expenditure was in any case on wages and salaries), any deflation of the 1935 National Income would increase the percentage represented by our estimate of R&D expenditure.

A country which today devoted 0.6 per cent of its National Income to R&D would find itself very much in the second rank. However, in the mid-1930s this represented a remarkable effort. At the same period the United States was only spending 0.35 per cent of its National Income on R&D.[72] Due to lack of data it has not been possible to make similar calculations for any other years. Nevertheless it is worth noting that the Soviet Union's expenditure on science in the year 1928/29 was 0.4 per cent of National Income for that year (in the United States in 1930 0.2 per cent was spent).[73] This suggests that in six years the Soviet Union increased the proportion spent on R&D by more than a half. But, as was noted above, in the second half of the 1930s National Income grew faster than real expenditure on science, and the 1940 expenditure of three milliard rubles, which would appear to include a wider range of expenditure than the figures for the earlier years, represented only about 0.45 per cent of National income.

However, we have, as yet, said nothing about productivity; the productivity of the Soviet research network was undoubtedly lower than that of western countries. The ability of the 'average' scientist was probably less than that of his counterparts abroad. The cut in the numbers of scientists in 1933 is clear evidence that over the previous years the demand for them had outdistanced the supply of suitable manpower. In fact, even in 1930 there had been talk of a need to slow down the rate of expansion of research establishments because of a shortage of scientists.[74] Research manpower was young and inexperienced. In 1931 one third of a sample of the research establishments' scientists were under thirty and three-quarters under forty.[75] In a 1934 survey only one third had had five or more years' research experience.[76] While nearly all the scientists had received higher education, many of those working in research in the 1930s had obtained theirs during the First Five Year Plan, when much higher education was sketchy.[77] Auxiliary and technical personnel were also often of a low standard.[78] As the 1930s progressed those scientists who had entered the research establishments in the hectic years of the early 1930s were obviously gaining experience. On the other hand at the end of the period remarks were still being made about the lack of experience of some R&D personnel[79] and even in 1939 a relatively large proportion of the country's scientists were under thirty.[80]

Some institutes, such as TsAGI or the Leningrad Physical Technical Institute, were as well-equipped as comparable establishments abroad.[81] However, even the majority of the Soviet Union's industrial research establishments, which were generally considered to be in a much better position than those in other fields,[82] were relatively ill-equipped. There were persistent shortages of apparatus and materials. In 1927 the demand for chemical reagents was only half satisfied, even after imports,[83] and while supply began to catch up with demand in the second half of the 1930s, there were still shortages at the end of these years.[84] An integral and vital part of every institute was the workshop which manufactured the small conventional pieces of equipment which could easily be purchased in other countries.[85] Several establishments had gone a stage further and were producing for sale equipment which was often otherwise unobtainable without using precious foreign currency.[86]

The purges of the late 1930s were to have a devastating effect on the performance of Soviet R&D. Scientists were arrested and sent to camps, where many famous scientists – N. I. Vavilov the geneticist is an example – were to die. The catastrophic effects that the mass arrests had on R&D led to the formation in the fields related to defence of prison research organisations – 'the first circle' of which Solzhenitsyn was later to write on the basis of his sojourn in such an establishment in the years after the Second World War.[87]

We have seen, therefore, that widespread reference by the leaders of the party and government to the importance of science was accompanied in the years before 1932 by a very rapid growth in the resources allocated to that field. This vast expansion was followed by a period of reorganisation and retrenchment which was linked to the general economic crisis at the end of the First Five Year Plan. The remainder of the 1930s was a time of slower growth during which expenditure on R&D may have declined as proportion of National Income. Over the whole period between 1923/24 and 1940 expenditure on 'science' grew by about thirty times in real terms and at its peak in the mid-1930s a remarkably high proportion of Soviet National Income was being spent on science. On the other hand the lack of the necessary highly trained manpower and continuing shortages of materials and equipment suggest that the productivity of the Soviet R&D network was for a long time relatively low.

3 The Industrial Research Effort

As we have seen, the new regime inherited few R&D organisations geared to the servicing of industry. The initial network of about a dozen industrial research organisations comprised the small number of bodies which existed before the revolution, such as the central laboratory of the War Department and the experimental factory of the War Chemical Committee, two previously private institutes which were nationalised and establishments organised on the basis of work previously done on a more informal footing in higher educational establishments and elsewhere by scientists such as Zhukovskii the aeronautical specialist. These institutes were to come under the control of VSNKh, the commissariat which had been set up after the revolution to run industry.[1] It was on the basis of such institutes that the Soviet industrial R&D effort was to grow. While in the West industrial R&D was mainly done in laboratories or departments attached to factories, in the Soviet Union such facilities were to play a minor role and R&D was to be concentrated in independent organisations. Two of the factors which helped to bring this about were the failure to develop research at enterprises before 1917 and the Soviet government's encouragement of the formation of large industrial research institutes, the benefits of which were to be particularly stressed with the introduction of the centrally directed industrialisation programme. These institutes were to combine the best features of the large industrial research laboratories of leading American firms with the advantages of the socialist organisation of production; for since there was no need for the duplication of research facilities which resulted from the secrecy surrounding R&D in the capitalist countries, in the Soviet Union there would be one large central research institute for each branch, or important sub-branch, of industry.[2]

By the mid-1920s there were thirteen institutes and laboratories[3] with a staff of 1000 under VSNKh (see Table 3.1). As in the case of the

The Industrial Research Effort   21

TABLE 3.1 Manpower of the USSR's industrial research establishments 1925-39 (beginning of year except where stated, thousands)

	1925	1928	1929	1930	1931	1932	1933 (i)	1933 (ii)	1934	1935	1936	1937	1938	1939
All industrial research establishments	—	—	—	—	23.6[cd5]	—	—	—	38.9[c8]	44.0[bf10]	50.3[g13]	—	—	35.2[17]
of which: scientists	—	—	3.4[bc3]	—	9.1[cd5]	(16)[6]	—	(20)[6]	(16)[6]	12.3[c11]	—	—	—	12.5[17]
Research establishments of VSNKh SSSR (-1932) and NKTP SSSR (1932-1939)	1.0[1]	4.0[a2]	4.6[4]	11.6[4]	16.9[4]	31.9[7]	45.2[e7]	—	34.3[9]	34.3[b10]	35.2[a14]	29.3[h15]	>20[j16]	>20[j18]
of which: scientists	—	2.0[a2]	2.4[4]	4.9[4]	6.3[4]	9.2[7]	13.3[e7]	—	9.4[9]	9.6[c12]	—	7.2[h15]	6.0[j16]	—

The figures enclosed by ( ) are estimates

Notes

a   1 October.
b   1 April.
c   Survey data, some establishments not included.
d   1 March.
e   Plan figure.
f   Excludes personnel under republican industrial commissariats, perhaps 2000–3000.
g   Average employment in March.
h   1 July.
j   Date unspecified.

See p. 22 for sources.

22 Science and Industrialisation in the USSR

SOURCES TO TABLE 3.1

1. *Nauchnye Dostizheniya v Promyshlennosti i Raboty Nauchno-Tekhnicheskogo Otdela VSNKh SSSR* (Moscow, 1925) p.34.
2. M. Ya. Lapirov-Skoblo, 'Nauchno-Issledovatel'skaya Rabota v Promyshlennosti', *NR*, no. 1 (1929) p.31.
3. *Nauchnye Kadry i Nauchno-Issledovatel'skie Uchrezhdeniya SSSR* (Moscow, 1930) p.18.
4. A. Ziskind, 'Nauchno-Issledovatel'skie Kadry Promyshlennosti', *FNIT*, no. 7-8 (1932) p.106.
5. *Narodnoe Khozyaistvo SSSR* (Moscow-Leningrad, 1932) p.551.
6. Estimates based on data for scientists and technical supporting personnel from *Sotsialisticheskoe Stroitel'stvo SSSR* (Moscow, 1935) p.625; it was assumed that in 1932 and 1933 the ratio between these two categories of staff was 4 scientists to 3 supporting personnel and for 1934 the ratio was taken from survey data (see source 8).
7. A. Ziskind, 'Kontrol'nye Tsifry Nauchno-Issledovatel'skikh Institutov Tyazheloi Promyshlennosti na 1933g., *SRIN*, no. 9-10 (1932) p.241; he gives a figure of 40.1 for employment on 1 July 1932, of which 11.2 were scientists.
8. *Kul'turnoe Stroitel'stvo SSSR v Tsifrakh: ot VI k VII S"ezdu Sovetov (1930-1934gg.)* (Moscow, 1935) p.148; data for 145 (out of 178) institutes, 54 (out of 82) branches.
9. N. I. Bukharin, 'Nauchno-Tekhnicheskoe Obsluzhivanie Promyshlennosti', *SRIN*, no. 3 (1934) p.6.
10. *Trud v SSSR* (Moscow, 1936) p.49.
11. *Kul'turnoe Stroitel'stvo SSSR. 1935* (Moscow, 1936) p.226; data for 144 (out of 163) institutes, 54 (out of 82) branches; the figure given for total manpower is 31,600.
12. *Ibid.*, pp.230-1; data for 95 (out of 108) institutes, 25 (out of 27) branches.
13. *Chislennost' i Zarabotnaya Plata Rabochikh i Sluzhashchikh v SSSR* (Moscow, 1936) pp.215-19, 236-45.
14. *Ibid.*, p.215.
15. L. Reinberg, 'Pokonchit' s Otstavaniem Nauchno-Issledovatel'skikh Institutov Promyshlennosti', *FNIT*, no. 10 (1936) p.101.
16. K. Novikov, *Industriya* 26 April 1938.
17. *Kul'turnoe Stroitel'stvo SSSR* (Moscow, 1940) p.238.
18. B. Volov, *Industriya* 11 May 1939.

---

research effort as a whole, the following years of recovery and growing capital investment saw a rising commitment to such research organisations. Spending on industrial institutes grew, perhaps, more slowly than total expenditure on science before 1928 but it was still about four times greater than in 1923/24 in real terms (see Table 3.2). However, in 1928, as a result of a resolution of the Fifteenth Party Congress, the Central Control Commission undertook its review of the industrial research network and on considering its report Sovnarkom USSR passed a lengthy decree which envisaged large increases in manpower and funds.[4] Subsequently the network of such institutes grew much faster than the rest of the research establishments. The growth was in fact staggering. Between 1928 and the beginning of 1933 the number of industrial research institutes grew from 24 to 238, their branches from 8 to 164,[5] and their manpower probably by over ten times (see Table 3.1). In conjunction with the general reorganisation at the beginning of 1933, 44 institutes would appear to have been closed, the number of branches halved. Real expenditure may have actually fallen in 1934. There was to be a further radical review of a large part of the network of the industrial research institutes in 1936 when over 25 institutes and branches of NKTP, the commissariat for heavy industry, were either liquidated, amalgamated with other organisations – in some cases higher educational establishments in the field – or lost their independence by having their staff and equipment moved to the laboratory of an industrial plant.[6] Just as the total number of research institutes was to decline in the second half of the 1930s, so in 1939 there were only 146 industrial research institutes compared with the

TABLE 3.2 Expenditure on industrial research establishments in the USSR, 1923/24–1937 (million rubles)

	Budgetary expenditure (current prices)	Non-budgetary expenditure (current prices)	Total expenditure (current prices)	Index at 1926/27 prices (1927/28=100)
1923/24	5.8	0.6	6.4	24
1924/25	6.4	1.4	7.8	27
1925/26	10.7	3.1	13.8	43
1926/27	18.2	5.2	23.4	69
1927/28	21.5	14.8	36.3	100
1928/29	31.0	27.0	58.0	147
1929/30	(70)	(60)	(130)	(300)
1931*	n.a.	n.a.	(210)	(419)
1932	n.a.	n.a.	(260)	(429)
1933	88	212	300	458
1934	70	260	330	437
1935	89	n.a.	n.a.	n.a.
1936	n.a.	n.a.	n.a.	n.a.
1937	85	n.a.	n.a.	n.a.

n.a. = not available
Figures enclosed by ( ) are estimates with substantial risk of error.
* In 1930 a special budget was drawn up for the period October to December to change the budgetary year to coincide with the calendar year.
SOURCE Appendix 1, Table A1.4.

194 that had survived the reorganisation at the beginning of 1933; there had been a similar fall in the number of branch institutes. Their total manpower had declined to 35,200 (see Table 3.1).

During the 1930s, however, the relative importance of the industrial research institutes in the Soviet R&D effort was to be reduced as the numbers of other independent R&D organisations increased. Firstly, there was a certain piecemeal creation of *independent design organisations* which were engaged in designing both products and processes. One or two such bodies had been set up in the mid-1920s, for example for designing equipment for the textile industry[7] and for boiler design.[8] By 1935 there were at least thirty such organisations[9] and their numbers continued to grow, but at the end of the period they probably still represented only a relatively small proportion of the R&D effort; in some fields, for example the aircraft industry,[10] they were to play an important role in R&D.

Secondly, some independent *experimental facilities* for development and testing were also established from the beginning of the 1930s. Information on the number of such organisations is scarce, but in 1935 there were, under NKTP, eighteen experimental factories, two experimental installations and four experimental mines. These plants were concentrated in fields to which the government attached significant importance. The mines, for example, were concerned with the underground gasification of coal, a long-established priority research project in the USSR.[11]

Finally, mention must be made of the *project organisations* which were responsible for planning and designing new plants and major reconstructions of existing factories. Although, perhaps, only a small part of their work was R&D, which was, indeed, a subject for criticism,[12] they played a most important role in technological innovation. It was through these bodies that the large-scale introduction and diffusion of foreign technology took place. The first project organisations were the State Institute for Projecting New Metallurgical Factories, Gipromez, set up in 1926[13] with responsibility for iron and steel and engineering plants and the Chemical Equipment Construction Company, Khimstroi, established in the following year for the chemical industry.[14] Their numbers grew quickly in conjunction with the industrial construction programme of the First Five Year Plan. By 1935 there was one such organisation for nearly all the branches and important sub-branches of industry under NKTP – 29 in all.[15] In 1938 there were some sixty such bodies under NKTP[16] and similar organisations existed in the engineering industry, which had been separated off into a separate commissariat in 1937. The project organisations were also very large, The State Institute for Projecting Chemical Factories, for example, is reported in the 1930s to have had a staff of 1200 at its main office in Leningrad and two branch offices with another 400 personnel in each.[17] With their growing specialisation on a narrower area of industry, they were on average not as large later in the 1930s, but, nevertheless, the total employment in NKTP's project organisations in 1938 was in the region of 30,000.[18]

THE HIGHER EDUCATIONAL ESTABLISHMENTS AND INDUSTRIAL RESEARCH

With a total of only four million rubles spent through the budget on all research in universities and technical institutes in 1935 and with non-budgetary funds probably not available in any substantial amount, research of relevance to industry undertaken in the higher education sector can only have comprised a tiny proportion of total expenditure on industrial research. However, the growing stress on the research role of the educational establishments which was reflected in the Sovnarkom decree of November 1937 and which also included some strengthening of research at technical institutes at the expense of the research institute network[19] did apparently result in increasing research in industrial educational institutes. In 1938 40 million rubles was spent on research in the technical institutes under NKTP[20] and 6 million on those under the commissariat for engineering;[21] these figures would also exclude any costing of the time spent by the institutes' full-time teaching staff on research. The available evidence suggests that expenditure in 1940 would have been somewhat larger.[22] However, disquiet remained about the lack of research in the institutes[23] and they could not have been responsible for anything more than a small part of industrial R&D.

RESEARCH AT THE FACTORY[24]

During the economic difficulties of the immediate post-revolutionary years the laboratory facilities of factories were neglected in the struggle to maintain at least some production. With the recovery of industry more attention began to be paid to the role of science in production. This was primarily a result of the recognition of factory laboratories' vital role in production control and testing. However, from the beginning of the 1930s the stress on improving the scientific control of production was accompanied by an increasing stress on developing the laboratories' R&D activity. On the other hand factory managements were loath to spend funds on R&D which contributed nothing to current operational capability and thus nothing to plan fulfilment. Thus the factory laboratory continued to be looked on as being solely for the purposes of production control. In addition most laboratories were very poorly equipped and their staff were much less qualified than that of the research institutes. Nevertheless, the picture was not uniformly bleak and there were plants, such as the Svetlana electrical-engineering works in Leningrad, which played a considerable role in the R&D activity of their particular branch of industry. In heavy industry as a whole it is reported that in 1938 the central laboratories of enterprises were undertaking 40 million rubles of research,[25] which was the same sum as the educational establishments of that area of industry spent in the same year.

Two other factory bodies were engaged in work classifiable as R&D. A small but growing amount of experimental design work was being undertaken by individual factories in their own design bureaux. The Stalin car factory, for example, seems to have tackled the design, manufacture and testing of the prototypes of the lorries ZIS-5 and ZIS-6 and the bus ZIS-8 in the years 1931–3,[26] and the Kirov factory in Leningrad (the pre-revolutionary Putilov works) was engaged in tractor and tank design in the 1930s.[27] In the machine tool industry the factory design bureaux represented the majority of the industry's design force.[28] This branch was, however, very much an exception within the engineering industry, in which the factories were generally critcised for their lack of design activity.[29] Some factories are also reported to have had experimental shops where new products and processes were developed.[30] Little is known about their work but, due to the great emphasis on gross output, it seems probable that in most cases these shops would have undertaken the batch production of the product or process which they had developed.

THE ACADEMY OF SCIENCES AND INDUSTRIAL RESEARCH

A particular feature of the Soviet R&D system was the gradual emergence over the years after the revolution of the Academy of

Sciences as an important source of R&D activity. Before 1917 the Imperial Academy had not been concerned with the technical sciences, and notwithstanding pressure from some scientists[31] the Academy was not to be involved with these disciplines until the late 1920s. Then an important corollary of the Soviet government's moves to create a more sympathetic Academy ideologically was an emphasis on the link between science and the economic life of the country.[32] An increase in the number of academicians, which was one of the measures which the government forced on the Academy, was to include the election of the first representatives of the technical sciences. Four were elected at this time. Their number was increased to eighteen in 1932 and in 1940 twenty-eight out of the total of 120 academicians were representatives of the technical sciences.[33]

The Academy's first research institute for the applied sciences was the Power Institute, which was founded in 1931,[34] and in 1932 111 out of the total of 1642 in its central research establishments were working in the technical sciences.[35] The next step in developing the role of the technical sciences in the Academy was the establishment in 1932 of a technical group in the division of the Academy for physics and mathematics; in 1935 this became a division in its own right.[36] There was, however, no large growth in the numbers of technical research institutes as an immediate result of these developments; it was not until the late 1930s that any substantial expansion in the research base of the Academy in this field occurred, when in 1938 five new institutes were added to the two institutes which had been set up previously for the applied sciences.[37] In terms of resources therefore the Academy's contribution to R&D of interest to industry was small and it was also only a small part of the Academy's total scientific effort. However, by the outbreak of the Second World War the Academy, as we shall see below, had an important role to play in the overall coordination of applied research.

THE RELATIVE SIZE OF SOVIET INDUSTRIAL R&D

It is clearly impossible to produce a series of reliable global figures for the number of people employed in industrial R&D and for expenditure on it. Some idea of its relative size, however, can be gained by comparing the firm figures for employment in the industrial research establishments with available data for employment in the research laboratories of industrial companies in the United States. As a result of the combined effects of the Soviet Union's efforts to expand R&D and the depression in the United States, we find that in the mid-1930s there were probably more people employed in research establishments in the Soviet Union than in the laboratories of American firms. The manpower of American laboratories was estimated to be 20,000 in 1927, five times the 1928 level in the Soviet Union; by 1931 it had grown to 50,000, but was then only 50 per cent more than the Soviet figure. In the following three years it fell by 5000, so that a 1934 figure of 30,000 is compared with a Soviet figure of around 40,000. However, in the following years, while employment in the industrial research establishments in the Soviet Union appears to have grown little, if at all, employment in the United States began to increase again and in 1938 it was over 50 per cent greater than in 1934.[38] The proportion of scientific (as compared with technical and auxiliary) personnel was considerably higher in the United States,[39] and it would seem probable that the numbers engaged in R&D, mainly Development, in other departments in American industrial firms outnumbered R&D personnel outside the Soviet Union's industrial research institutes, although the gap may have narrowed in the second half of the 1930s.

A notable difference between the industrial research networks of the Soviet Union and the United States was the average size of the research establishments. In 1935 average employment in the Soviet industrial research establishments was 190, with around 50 scientists in each establishment.[40] The available data for the United States suggests that the average industrial company which undertook research employed under twenty research personnel in 1933, around twenty-five in 1938.[41] However, the United States figure did not reflect a uniform smallness in research effort. In 1938 fifty-four companies (around 30 per cent of the total undertaking R&D) employed over half of the total research personnel – an average of 450 per company.[42] Unfortunately this cannot be taken as a true reflection of the size of the industrial research laboratories since we do not know how many independent laboratory organisations were under each of these companies.[43] Nevertheless, it is clear that, while the average industrial research organisation in the United States was much smaller than its Soviet counterpart, the former's largest laboratories were comparable in size with the Soviet Union's large industrial research institutes.

THE GEOGRAPHICAL DISTRIBUTION OF INDUSTRIAL RESEARCH

The debate over the Soviet Union's industrialisation programme was marked by intense discussion of the location of new industrial development; whether new plants should be constructed in existing industrial areas or established further to the east in the Urals and Siberia, strategically safer areas where industry was underdeveloped and resources virtually untapped. In spite of some strong, particularly Ukrainian, opposition the decision was made to undertake an enormous industrial expansion in these regions. The years of the First Five Year Plan saw the start of work on the Urals-Kuzbass combine, the key to much of the planned expansion, a huge iron and steel complex which was to be based on the Urals iron-ore deposits and the coal resources of the Kuzbass. New engineering plants were to follow.[44]

A corollary of this industrial policy was a stress on the need to pay similar attention to expanding research facilities in the areas of new industrial development and away from Moscow and Leningrad, the existing main scientific centres. The Sovnarkom decree of August 1928 pointed out that sufficient attention had not been paid to the organisation and expansion of research in the country's industrial regions. VSNKh was to take this into account when planning the future development of industrial research. In stimulating research in these areas it was to make full use of the research facilities of provincial educational establishments and of local research organisations under Narkompros and other departments. But the decree envisaged in particular the foundation by existing institutes of branches in the important provincial areas.[45] Consequently in July 1929 VSNKh amended the standard regulations for the organisation of research institutes by including a section specifically dealing with the formation of branches.[46]

The subsequent rapid expansion in institutes' branches was a clear reflection of this policy. Those attached to VSNKh research institutes grew from eight in 1928 to ninety-five at the beginning of 1932.[47] However, such branches and also newly established independent institutes in the provinces were generally poorly staffed.[48] Indeed these developments appear to have had only a marginal effect on the concentration of industrial research scientists in the main cities of Moscow and Leningrad. In 1929 just under eighty per cent of these scientists worked in these two cities and at the beginning of 1932 the proportion was still over seventy-seven per cent, well over twice the proportion of the industrial labour force which was employed in these two centres.[49]

The planning documents issued in connection with the drawing up of the 1932 plan for research in VSNKh envisaged further regional expansion[50] and at the Seventeenth Party Conference in February 1932 Bukharin, then the head of research in VSNKh, spoke of the need to press on with the decentralisation of industrial research.[51] In 1932 the rate of growth of new institutes even accelerated. In industry as a whole there were 119 branches attached to existing institutes at the beginning of the year and 164 at the end.[52] This may have resulted in a fall in the proportion of industrial research scientists who worked in Moscow and Leningrad. On the other hand the reorganisation at the beginning of 1933 which resulted in a halving of the number of branches but only a twenty per cent cut in institutes can be presumed to have had a much greater effect in the provinces on the numbers working in research establishments. In 1935 just under three quarters of industrial research scientists were in Moscow and Leningrad and, in spite of the continuing and increasing industrial development away from existing centres, the share of these two cities was slightly greater at the end of the decade.[53]

Thus, while we have no information on the changes in the distribution of total manpower in industrial research, it is clear that in spite of the adoption of a policy of encouraging the expansion of research in the provinces, the vast majority of scientists continued to work in either Moscow or Leningrad. Nevertheless, in 1940 there were at least 1195 scientists working in industrial research in the RSFSR outside these two centres, well over four times the number in 1929, and the number of scientists in the other republics had increased by a similar magnitude.[54]

The geographical distribution of the personnel working in other independent R&D establishments is unknown. However, the available evidence suggests that these organisations were also concentrated on Moscow and Leningrad. In 1935 all of the design organisations for which the information is given (eleven out of thirteen) were in these two cities, as were half of the experimental facilities and at least twenty-six out of the twenty-nine project organisations.[55] Factory laboratories were undoubtedly more widely dispersed but some of those in newly developing areas were extremely poorly staffed and equipped.[56] It would, therefore, seem likely that industrial R&D manpower in the Soviet Union was substantially concentrated throughout the period in its two major cities.

THE WORK OF THE INDUSTRIAL RESEARCH NETWORK

When a country is importing a large amount of foreign technology, it seems reasonable to expect that a large part of the activity of industrial research facilities will be directed towards assisting in the introduction of this technology into domestic industry. This introduction can take two main forms. Firstly, research organisations can undertake the testing of different processes and products to ensure that the most suitable are chosen – those corresponding best to the needs of the country and its available resources. Secondly, they can ensure that innovation will be successful by adjusting the imported technology to the recipient country's different conditions (geographic, climatic or socio-economic), or the different nature of its raw materials. Such activity will range from minor adaptations undertaken in the factory itself to major modifications requiring substantial applied research.

This was indeed true of the Soviet Union in this period. The initial major task of its first research institute for the study of machine-tools was to assess the various foreign makes and models, and to select the most suitable range for domestic production.[57] The Auto-Tractor Research Institute tested different makes of engines,[58] the All-Union Electrical-Engineering Institute foreign electrical equipment.[59]

The industrial research institutes were also adjusting foreign technology to Soviet materials. This was clearly of prime importance when the country was making for the first time products in which the nature of the raw materials is crucial, for example, refractory materials.[60] This task included not only ironing out the difficulties caused by the difference between Soviet and foreign deposits of the same materials, but also work directed towards substituting materials available in the Soviet Union for those which had to be imported. Thus, the Central Institute of Metals investigated the possibility of substituting chrome and other metals for wolfram in the production of high-speed steels for the metal fabricating industry.[61] Analogous efforts were made to find substitutes for those materials which were in short supply – 'deficit' materials. Plastics were seen as a possible alternative to some non-ferrous metals,[62] and there is a report of research into the substitution of iron for copper in the electrical industry.[63]

Table 3.3 shows the breakdown of manpower in the Soviet Union's industrial research institutes and in the research laboratories of American industrial companies for various years in this period. In spite of the anomalies – for example, the great variations in the percentages for chemicals and metallurgy in the Soviet Union – these data would appear to support the view that a large proportion of the work of the industrial research institutes was concerned with the utilisation of existing foreign technology. For it seems that in the Soviet Union a much smaller percentage (perhaps only half that of the United States) were working in the chemical and electrical industries – the science-based industries – in which R&D was of the greatest importance at the time. However, in view of the neglect of the chemical industry in the Soviet Union, this argument cannot be considered conclusive.

The Soviet Union could not rely entirely on foreign technology. The military field immediately suggests itself as an area in which one would

TABLE 3.3 Breakdown of manpower employed in the Soviet Union's industrial research institutes, 1929, 1931 and 1935, and in the research laboratories of United States industrial companies, 1927 and 1938, by field of research institute in the USSR, of company in the United States (in percentages)

	USSR			USA	
	1929[a]	1931	1935	1927	1938
Metallurgy	24.3	20.2	9.2	6.4	6.2
Electrical industry	16.8	11.6	8.8	31.6	19.8
Power			7.1	2.8	2.2
Chemicals	15.0	27.3	19.8	28.4[b]	38.1[b]
Fuel and Mining	11.2[c]	15.2[c]	19.9[c]	–	–
Engineering	8.5	6.6	10.6	12.7	14.0
Food, Leather, Textiles, Forest and Paper Products	6.3	–	6.7	4.6	4.9
Others	17.9[d]	19.2[d]	17.9[d]	13.5[e]	14.8[e]
	100.0	100.0	100.0	100.0	100.0

– not given

NOTES
[a] Breakdown for scientists only.
[b] Figures include manpower employed in companies engaged in the exploration and production of petroleum; this appears under Fuel and Mining in the USSR.
[c] Includes those engaged in the field of mining machinery and equipment.
[d] Includes construction (about ten per cent) which is not covered by the US data.
[e] Includes undistributed manpower in consulting institutes and research associations.

SOURCES
USSR: 1929: *Nauchnye Kadry . . .*, pp. 20, 77.
1931: *Narodnoe Khozyaistvo SSSR* (Moscow, 1932) pp. 549–51.
1935: *Kul'turnoe Stroitel'stvo SSSR. 1935*, pp. 227–9.
USA: Perazich and Field, *Industrial Research . . .*, pp. 61–2. 69–71, 77.

expect it to have been very difficult to obtain technical assistance. However, for part of the period the Soviet and German governments had secret agreements for such help.[64] The German influence on Soviet military technology was undoubtedly substantial, but not all-embracing. In the field of aircraft design, for example, the Soviet Union was to a great extent self-dependent. There is also a report of great difficulties in acquiring technical help in the chemical industry, which were partly due to strategic considerations.[65]

In areas where no suitable technology existed, the Soviet Union also had to lean on the abilities of its own R&D organisations. In some cases this was a further result of the differing characteristics of raw materials; thus, it was stated that while in the period of the First Five Year Plan use was made of technical assistance in the processing of the country's peat, the nature of the actual deposits was so different that extraction technology had to be developed independently.[66] In other cases the Soviet government embarked on the development of technology in fields which market economies had declared unprofitable. Strategic reasons drove it to fund the development of artificial rubber,[67] and a desire to save imports led to the use of low-grade deposits of the Kola peninsula for superphosphate production.[68]

In spite of such examples it is undeniable that innovation of processes and products developed in the Soviet Union was relatively small, and concentrated in areas to which special priority had been given. Its level, however, would not seem to reflect a lack of suitable results from applied research so much as a failure to get the full return from this research in the form of products and processes introduced into production.[69]

Despite the needs of the industrialisation drive, there was not a complete neglect of basic research, which can be looked on as providing for 'tomorrow's' technology. Among the industrial research institutes were several engaged to a large extent on fundamental research.[70] By the mid-1930s these had been specially designated as 'head' institutes and were seen as having the specific task of solving fundamental theoretical questions and seeking out the main directions for future research. They included such leading centres for basic research as Ioffe's Leningrad Physical Technical Institute and its offshoot the Ukrainian Physical Technical Institute – two of the major nuclear research establishments – and the Karpov Chemical Institute. An English chemist working in another, the Institute for Chemical Physics in Leningrad, apparently found greater opportunities for basic research than in Britain.[71] Later in the 1930s, as the administrative structure became more narrowly specialised, such institutes passed into the control of the Academy of Sciences.

The industrial R&D organisations were also doing other types of work than that linked to the import of foreign, or the development of domestic technology. This included much routine testing and standardisation work on domestic products and materials. J. G. Crowther wrote of the State Ceramic Institute at the start of the 1930s that its work in this field was comparable to that done by the American Bureau of Standards.[72] TsAGI, the main institute for the aircraft industry, in the mid-1930s was regularly engaged on testing the parts produced in the aircraft factories and did all the control testing for factories which lacked the necessary laboratories. It is even stated to have used its development facilities for batch-producing parts which were technically beyond the capability of the aviation factories.[73]

Several institutes were manufacturing scientific equipment. The Ukrainian Physical Technical Institute, in particular, appears to have produced a large range for sale to other institutes and to factory laboratories[74] and Gintsvetmet, the main institute for non-ferrous metallurgy, produced 'nearly all' the specialised laboratory equipment for that branch of industry.[75] Similarly, the State Institute for Applied Chemistry and the Institute for Chemical Reagents produced chemical products for sale.[76] The latter, in particular, was an important producer of high-quality reagents.

Among the research institutes under the industrial commissariats were some whose main purpose was to undertake geological survey work. The most notable of these, the Central Geological Survey Institute, had two branches and a total staff of 1300 in the middle of the 1930s.[77] The size of such survey work was probably reflected in the proportion of personnel in Mining and Fuel in Table 3.3. Other institutes, for example the Research Institute for Fertilisers, were also carrying out some surveying work.[78]

Another important function of the institutes was the training of future research personnel. In 1929 about 400 and in 1939 about 500 postgraduate trainee researchers were studying in the industrial research institutes.[79]

THE R&D CONTENT OF THE WORK OF THE INDUSTRIAL RESEARCH INSTITUTES

Routine testing and standardisation work, geological surveys and manpower training are not R&D in the Frascati sense;[80] nor, of course, is equipment manufacture. This applies similarly to the testing of foreign equipment undertaken by the research institutes in connection with the utilisation of imported technology. However, work on the adjustment of foreign technology to Soviet conditions during the period of introduction is harder to classify. Here it is difficult to draw the boundary between R&D and non-R&D activity. The Frascati manual gives no recommendations for the classification of work linked to technological imports. It seems more suitable to follow a United Nations Industrial Development Organisation monograph[81] and to differentiate between *adaptation*, in which 'the essential elements of imported production processes ... remain unchanged' and *modification*, requiring 'some original thought'.[82] The clearest instance of the distinction between these two categories is the difference between the *adaptation* of a product or process to the importing country's deposits of the necessary raw materials, and the *modification* of a product or process in order to use different materials. The work of the Central Institute of Metals on finding substitutes for wolfram to which we have just referred is, therefore, an example of modification.

The Soviet Union's industrial research institutes would appear, in general, to have been working to a much greater extent on modification than on adaptation and thus the majority of their work on adjusting foreign technology can be considered as R&D.[83] Adaptation was probably mainly undertaken by the factories themslves during the commissioning period. Nevertheless, such was the size of routine testing, geological surveys and other non-R&D activities that throughout the period a substantial amount of the activity of the staff of the industrial research establishments was not strictly R&D.

Thus, in the period under review industrial R&D was concentrated, although to a decreasing extent as the 1930s progressed, in specialised research institutes under the commissariats responsible for industry. While in 1917 the new regime inherited little in the way of industrial research facilities, by the end of the First Five Year Plan a large network of these institutes had been established to service all branches of industry. Notwithstanding subsequent retrenchment their manpower grew from 1000 in 1924 to 35,000 in 1939. Additionally, from the first half of the 1930s there emerged a growing number of independent design and development organisations. Relatively little research was done by the factories themselves, by higher educational establishments or at the Academy of Sciences. Despite a frequently stated policy of geographical dispersal the industrial R&D force was largely concentrated in Moscow and Leningrad. The work of the industrial R&D organisations was aimed both at enabling the Soviet Union to make the fullest use of imported technology and at producing new indigenous technical developments. Their activities ranged from theoretical research to routine testing and the commercial manufacture of scientific equipment. Consequently probably a substantial proportion of their activities was not strictly classifiable as R&D.

4 The Central Control of Industrial Research, 1917–30

The Imperial Academy of Sciences and the scientific establishments which before the revolution had been under the Ministry of Education became the responsibility of the Narkompros RSFSR, the People's Commissariat of Education of the newly established RSFSR (which comprised most of what was in 1923 to become the USSR); a scientific department was established as one of its seventeen departments.[1] On the other hand the organisations with a military and industrial orientation, such as the Central Laboratory of the War Department, which had been set up just before or during the war under or closely linked to the War Department, came under the control of the commissariat set up to control industry, VSNKh, the Supreme Council of the National Economy. The latter, thus, presented a potential challenger to Narkompros's inherited primacy in science. Indeed, by the beginning of 1918, there was talk of an independent body for coordinating science.[2] Lenin apparently wanted to organise a commissariat for science and technology.[3] N. P. Gorbunov, a young protégé of Lenin and secretary of Sovnarkom RSFSR (the body which comprised all the commissars and was responsible for the overall control of the economy), was entrusted early in 1918 with drawing up the statute for this body. He soon found that he had to contend with a widespread reluctance to see any importance whatever in the project[4] and he also encountered bitter opposition from Narkompros, which saw that its position in science was being very definitely threatened. To retain its dominance, Narkompros proposed in the spring of 1918 that a Russian Association of Sciences should be established under Narkompros to unite all research establishments in the republic.[5] There was in addition growing friction between Narkompros and VSNKh over the control of the growing number of new research establishments which were being created.[6] The Academy, too, became caught up in this rivalry. Lenin's close interest in science ensured that from the outset the Academy not only had dealings with Narkompros but also some direct contact with Sovnarkom, of which Lenin was chairman.[7] In fact, the Academy presented its first series of financial estimates directly to Sovnarkom in April 1918.[8] To make matters worse from the point of view of Narkompros, Sovnarkom then redirected them to VSNKh. Narkompros did finally receive these estimates of the Academy and formally transmitted them to Sovnarkom, but there was obviously a danger that the Academy might be transferred to VSNKh. Indeed Narkompros was already incurring Lenin's displeasure for the way in which certain of its leading members, including the Marxist historian M. N. Pokrovskii, who was deputy commissar of education from 1918 to 1932 and particularly involved in higher education and scientific research, had adopted the proposal for a Russian Association of Sciences as a means of doing away with the Academy in its existing form.[9] Further, in June 1918, Gorbunov wrote a letter to the Central Committee of the Party criticising Narkompros's activity in scientific matters and calling for the transfer of the Academy and other research establishments to the control of 'a new scientific organ' under VSNKh.[10]

However, Gorbunov's proposals for this body – for it was as such that Lenin's Commissariat was now viewed – were themselves reformulated by several redrafting commissions which took into account demands by Narkompros that changes should be introduced which 'would leave Narkompros the primary position in matters of science'.[11] The outcome was the Sovnarkom decree of 16 August 1918 which established NTO, the scientific and technical department of VSNKh; despite Gorbunov's desire for a body controlling all research establishments, which he had expressed in his letter to the Central Committee, its role was to be limited to the applied sciences, although within this field it was intended to have wide-ranging functions in controlling development as well as research with responsibility for the central direction of applied R&D in the RSFSR and for providing advisory services on matters concerning science and technology.[12] The only remnant of its originally intended status as a commissariat was a paragraph in its statute which gave it the right to deal directly with Sovnarkom; this anomaly was removed by an amendment to the statute a year later.[13] The limitation of NTO's functions to applied research was, in a way, a partial success for Narkompros in defending its prerogatives in scientific research. However, the formation of this new organisation for the control of science as part of VSNKh and not as a commissariat in its own right is also said to be not so much a result of pressure from Narkompros as of the views of Gorbunov himself on the need for close links between science and industry.[14] An outcome of these events was that even from this early time the institutional framework was established for the development of two distinct networks of research facilities, one of institutes doing the 'pure' research in science under Narkompros, the other of institutes mainly doing applied research attached to the commissariats responsible for individual branches of the economy, mainly VSNKh, but also for example NKPS, NKZem and NKZdrav, the commissariats for communications, agriculture and health respectively.

For some years after the revolution the picture was more complicated. The new VSNKh department did not immediately take over control of all the institutes engaged in applied research of interest to industry. Institutes which had developed out of the work of the Academy's commission KEPS tended to fall within the purview of Narkompros in spite of the fact that they were engaged on applied research; the State Ceramic Institute, in fact, was to remain under its charge until 1926.[15] Indeed at the outset NTO did not even assume direct control of all VSNKh's scientific activity. Some institutes remained under the organs which were responsible for the individual branches of industry. Over the following years they gradually passed under the control of NTO, although it was not until 1924 that it achieved control of all the commissariat's institutes.[16] Initially, before the growth of its own facilities, NTO proposed and financed a considerable amount of work which was done by scientists in establishments and laboratories which were not attached to it.[17] Similarly it was not until 1921 that all the various technical advisory services which had been established for industry were centralised under the Central Scientific and Technical Council of NTO.[18]

Scientific and technical departments on similar lines to NTO VSNKh RSFSR may also have been formed under the Supreme Councils of the National Economy of the other republics. Certainly the Ukraine had such a department from June 1920;[19] it was apparently based on divisions of NTO which had been set up in Kiev and Khar'kov in 1919.[20] There was also, after the transfer of the seat of government to Moscow, a short-lived Petrograd division of NTO under the Council of the National Economy of the Northern Oblast – a recognition of its importance as a scientific centre.[21]

During the first years of NTO's existence the expansion of research had undoubtedly been restricted by the economic circumstances. At the end of 1923 it controlled twelve research institutes of which ten had been founded in the years 1918 and 1919. Nevertheless it had begun to develop the auxiliary organisations which provided services necessary for research – publishing houses, establishments which made laboratory equipment and apparatus.[22] It was, however coming under criticism for its failure to play a sufficiently positive role in developing R&D.[23] Its work was handicapped by the fact that of the original board of four (in 1920 it was increased to eight)[24] appointed to run it only Gorbunov was not a working scientist. The others were too occupied in teaching and research to devote much time to NTO business.[25] Moreover, for two years from the time in 1920 when Gorbunov was appointed administrator of Sovnarkom, it seems to have lacked a recognised head. NTO also had no direct access to the Presidium of VSNKh. It was not until 1923 that the head of NTO was to sit on the Presidium as of right.[26]

Thus, during the early years of the new regime, the government after some debate established a central body to control industrial research under the major government organ responsible for industry. During these same years the government established central control over the whole of industry. Indeed during the Civil War centralisation was extreme. From this time forward a continuing source of major debate among all connected with the administration of Soviet industry was the optimum degree of centralisation. With the advent of the New Economic Policy in 1921, many of the decisions previously taken by VSNKh itself became the province of the enterprises and the trusts into which enterprises were grouped. Any talk of a return to more central control provoked suggestions of a return to 'glavkism' – VSNKh's *glavki* (chief administrations) for each branch of industry had been the major instrument of the centralisation of the Civil War years – and of complaints about bureaucracy.[27] The following years were to be marked by a continuing search in changing economic conditions for a practical form of combining administrative decentralisation and the delegation of authority with the overall central direction of the process of industrialisation. Changes in the administrative structure for industrial research were closely connected with the developments which attempted to solve this problem. The major alterations in the administrative system for research were not so much a result of a specific review of the problems of controlling an R&D network as a by-product of general debate and change in the administrative structure for the whole of industry. These changes were also to influence the wider question of the organisation of overall national science policy-making.

The Central Control of Industrial Research 41

In the years under discussion there were two distinct periods in the organisation of the industrial research establishments. Until the end of 1929, large-scale industry was controlled by one commissariat, VSNKh, and its research network by a single department of that commissariat, initially NTO and then after 1926 NTU, the Scientific and Technical Administration. The second period began with a reorganisation of the administration of industry in December 1929. The decentralisation in the control of industrial production which was introduced at that time was accompanied by a simultaneous decentralisation in the control of industrial research. After this time the industrial research network was never again to be subordinate to a single body.

In 1922 a radical change in the work of NTO occurred with the appointment of Academician V. N. Ipatieff to head the department. Ipatieff, a leading chemist and former artillery general, had done much practical work in the chemical industry both before and after 1917; in 1922 he was head of Glavkhim VSNKh's chief administration for the chemical industry and he moved to NTO on the dissolution of that body as part of the decentralisation measures inaugurated by the New Economic Policy. He was also a member of the Presidium of VSNKh. Ipatieff was probably the person in the Soviet Union who was best qualified to head NTO with its emphasis on the linking of science and industry. He soon began a review of the role of NTO in fostering research and of the activities of the research establishments. This resulted in a rationalisation of the research programmes being undertaken by various scientists in laboratories at higher educational establishments which NTO was financing and a transfer of much work to VSNKh research institutes. A below-standard research institute was disbanded and its personnel and resources absorbed by other establishments.[28] However, he was soon having to defend the very existence of the department. In 1923 the central apparatus of VSNKh – now **VSNKh SSSR**, for, following the formation of the USSR, the existing VSNKh for the Russian republic became the all-union commissariat VSNKh SSSR and responsible for industry of national importance – was reorganised to take account of the changes in its role in industry which were embodied in a recently approved decree which had finally established the legal status and functions of the new industrial trusts. In this connection Ipatieff made a report on the work of NTO to the Presidium of VSNKh. At this meeting there was opposition to the continued existence of NTO by Pyatakov, a deputy head of VSNKh who was to the left of the party and who became a member of Trotsky's faction in the industrialisation debate. He apparently had a very low opinion of the quality of the membership of the board at the head of NTO; and, reflecting the current trend for administrative decentralisation, he suggested that the control and finance of industrial research should be vested in the recently established trusts.[29] However, the Presidium of VSNKh approved a statute for the department which basically restated its original functions, while placing greater emphasis on its role of ensuring that the Soviet Union kept abreast of the latest developments in science and technology abroad and reorganising the institutional structure for the provision of technical advisory services by splitting up the Central Scientific and Technical Council into ten Scientific and Technical Councils (NTSy), which were to be supervised by the board of NTO.[30] While on this occasion Ipatieff and the defenders of NTO won the day, their respite was to be short, for the matter was to be reopened a year later. In November 1924 a special commission was attached to GEU, the Chief Economic Administration of VSNKh, with the task of considering the future of NTO. On this occasion Pyatakov himself seems to have been the prime mover behind the setting up of this commission.[31] Possibly he and others in favour of decentralising the administration of research hoped that the previous decision to retain NTO would be reversed by Dzerhinskii, who had taken over as head of the commissariat in the previous February. The work of the commission was accompanied by a debate in the VSNKh newspaper *Torgovo-Promyshlennaya Gazeta* with the arguments put for and against the central control of research. Opponents of NTO viewed it as a barrier between the research institutes and industry[32] or as an overloaded bureaucratic organisation.[33] Its defenders pointed out that to hand the institutes over to the trusts or enterprises would lead to a fatal limitation of their work to industry's short-term needs[34] and claimed that NTO's institutes were, in fact, more 'practical' than some of the laboratories of foreign industrial firms.[35]

Early in 1925 Dolgov, head of the commission, presented a report first to the board of GEU[36] and then to the Presidium of VSNKh itself.[37] The report took a position which was on the whole favourable.[38] Dolgov argued that applied science and technology could produce good results only where backed by widely based exploratory research; he defended the work done under NTO as of interest to the state as a whole and not merely particular branches of the economy; he rejected the proposal that the research establishments should be entirely dependent on enterprises or groups of enterprises. No changes were proposed in the structure of NTO or in its administrative relationship with the rest of VSNKh, though the report advocated stronger links between NTO and the Presidium of VSNKh and those parts of the apparatus concerned with the planning and administration of industry. The positive recommendations of the report of Dolgov's commission were that the work of the Scientific and Technical Councils should be strengthened, and that a Council of Assistance (*Sovet Sodeistviya*) and also a Technical Conference (*Tekhnicheskoe Soveshchanie*) should be established by each research institute. These were bodies aimed at increasing contacts between scientists and those working in industry.[39]

At the meeting of the Presidium of VSNKh, which took place on 4 February 1925, Pyatakov insisted that Dolgov's proposals did not avert the danger of a complete divorce of research from industry and he called for an organisational link between the research institutes and industry.[40] In the newspaper report of this meeting there is no explicit support for his attack, although Shein spoke of the inadequacy of the scientific and technical manpower in the factories and the isolation of such specialists as there were in industry from scientific ideas.[41]

In reply to Pyatakov, Ipatieff pointed out that the strength of NTO's connections with industry was shown by the total of more than a million rubles which they received direct from industrial bodies.[42] The most important support for the report came from Dzerzhinskii.[43] He considered that it was a good sign that industry was beginning to criticise the insufficiency of its links with NTO. The development of the finance of research by special funds, as spoken of by Ipatieff, showed that the link existed; what was needed was to improve it. Institutes 'cannot have in mind the interests of a particular enterprise, but must consider general goals, general objectives, objectives which are of a scientific rather than a purely practical nature'. Consequently he saw several dangers in subordinating institutes to factories or trusts. The scale of activity of the institutes would be lowered and the country's scientific manpower, which was in any case insufficient, would tend to be dispersed and scattered. It would also mean that the institutes would no longer be financed directly through the budget; in such a situation, he said, there would be no state control over the correct organisation and administration of such institutes. Furthermore, since there were no trusts which covered the whole of a given branch of industry, unless duplication of research facilities occurred part of a branch would not be supported by an institute. These were all arguments for the continuing existence of a central body, and an organisation such as NTO also made for easier popularisation of science and dissemination of information about the achievements of the research institutes. Dzerzhinskii considered that the required improvement of the links between research and industry could be provided for by the proposed Councils of Assistance and Technical Conferences; these would provide the necessary link without destroying 'that independence and that necessary initiative and freedom which scientific institutes must have'.

In view of the strong support for NTO by Dzerzhinskii, it is no surprise that the Presidium accepted Dolgov's report; the board of NTO was requested to work out detailed plans to carry out the proposals for the creation of the new advisory bodies.[44] NTO thus survived the assault and the principle of central control for industrial research was, in the light of Dzerzhinskii's views, if anything strengthened.[45]

A year later another review of the structure of VSNKh was undertaken, and the role of NTO was further discussed. This new review was the result of a growing preoccupation with the increase in centralisation and bureaucracy; for the changing industrial situation was beginning to show flaws in the structure for running industry which had been adopted in 1923. The increasing amount of industrial development was resulting in a considerable overlap between GEU, which was responsible for the general supervision and planning of industry, and the Central Administration for State Industry, TsUGProm, which comprised a series of directorates responsible for particular branches of industry. The main features of the proposed reorganisation were agreed in April 1926 before Dzerzhinskii's death. In particular they envisaged a fusion of GEU and TsUGProm to produce new *glavki*, which were to be directly under the Presidium of VSNKh and to be responsible for groups of industries.

However, the future of NTO was apparently still under review in August. The plans to set up *glavki* clearly strengthened the case for the decentralisation of the research institutes, and Flakserman of NTO specifically argued against such a move in an article published in the August issue of NTO's monthly journal.[46] He reiterated the arguments of the previous year: that the institutes did work that was of interest to more than one industrial body; that enterprises wanted to use the institutes for their daily production servicing needs. Whatever residual opposition to NTO this article may have indicated came to nothing. The attitude of Kuibyshev, who took over VSNKh on the death of Dzerzhinskii, to NTO and its research institutes initially seems to have differed little from that of his predecessor. It is significant that in continuing the cost-cutting campaign (the 'Regime of Economy')[47] started by Dzerzhinskii, he suggested some exemption for research establishments, instructing the commission for cutting staff 'to treat the scientific and technical establishments very carefully, bearing in mind their significance and importance'.[48]

When the reform of VSNKh was put into effect, the central body for research indeed remained. Now entitled the Scientific and Technical Administration, NTU, it had the status of a *glavk*. In addition to taking over the previous functions of NTO, NTU was to control two further organisations, the Central Committee for Water Conservancy (before the reorganisation under GEU) and the patent-registering body, the Committee for Invention Affairs (formerly an independent body attached to the Presidium of VSNKh).[49]

The reorganisation of VSNKh led to the cutting in size of its Presidium.[50] As a result NTU was deprived of the direct representation on that body which NTO had had through Ipatieff and then through Trotsky, who had been associated with the running of NTO from June 1925 when the direction (*rukovodstvo*) of NTO was one of the posts which he was given after his removal from the post of Commissar of War.[51] In February 1926 he became chairman of the board and the official head of NTO;[52] however he took little part in the actual decision-making of the board and Ipatieff continued to be *de facto* head of the organisation.[53] The consequence of the 1926 reorganisation was a return to the position which had existed before the appointment of Ipatieff, by which it was placed under the general supervision of a member of the presidium. However, in practice this did not involve any diminution of its status, as Kuibyshev himself took on responsibility for NTU.[54]

After 1926 the developments which were to lead to a policy of swift planned growth for the Soviet Union were increasingly to confront VSNKh with technological decisions connected with future industrial expansion. As a result its central scientific organ, now NTU, was no longer to continue primarily as a scientific organ, but became part of the controversy over how best to use all the country's scientific resources in ensuring that the technological development of the country proceeded along the best lines. The policy of linking science and industry by creating advisory and coordinating bodies such as the Councils of Assistance and the Technical Conferences, rather than by attaching the institutes to industrial organisations, was in time to be more and more strongly criticised.

The beginning of a new attitude can be seen in a decree approved in December 1926, which was entitled 'the industrialisation of the country and the role of the scientific and technical establishments'.[55] This decree reorganised the leadership of NTU. Trotsky and Ipatieff were no longer to be at its head.[56] Its new chairman was to be V. M. Sverdlov, previously head of the *glavk* for the mining and fuel industries.[57] The appointment of a man who was an industrial administrator and not a scientist can, perhaps, be taken as a portent of the way in which it was intended NTU should develop. Those who had assisted Ipatieff as vice-chairman, Martens and Flakserman, also lost their posts although both of them and Ipatieff himself were to sit on the enlarged board of thirty-nine which was nominally to control NTU. The role of NTU in advancing production was emphasised more strongly than before; while it was an organ for the development of scientific and technical ideas, at the same time it had 'to become a support for economic organisations in the maximum application of scientific achievements to the practical activity of production'. The NTSy were to play a more active role in the technical development of industry with each council to become the centre of scientific and technical ideas for its branch of industry. Special attention was to be paid to the study of foreign technical achievements and to the dissemination of information on them to industry.

The immediately following years saw a rapid expansion in industrial research. Over the next three years manpower and funds each increased by roughly three times. This growth was marked by a dramatic change in the relationship between the research establishments and industry as non-budgetary funds from industry expanded much more rapidly than central budget funds; in 1928/29 they were to have risen to a half of total funding.[58] The origin of this emphasis on finance by industry would appear to be the review of the 1926/27 budget by the budget commission of TsIK. It is reported to have adopted as a basis for the future financing of VSNKh's research establishments the principle that they should receive funds on the basis of 'payment for services', thus enabling cuts to be made in state budget finance for them.[59]

The expansion of research increased the demand for scientific personnel. At the same time, with the growing emphasis on industrialisation, increasing attention was being paid to the country's overall supply of scientific and technical manpower. This was reflected in the resolution of the Fifteenth Party Congress on the work of the Central Control Commission, which proposed a study of the utilisation of this manpower.[60] As a consequence of this resolution a survey of the industrial research establishments was undertaken in the first half of

1928 and a draft resolution based on it presented to Sovnarkom in June.[61] This survey was carried out at a time when, with major capital projects the subject of fevered discussion in numerous VSNKh departments and commissions,[62] a lack of NTU involvement – in spite of the proposed reorientation of its work at the end of 1926 – in the immediate problems of future industrial development appeared to be giving rise to growing anxiety. Indeed, even before the completion of the survey steps were taken to improve the links between science and technology and industry which eroded the position of NTU. In May 1928 the overall direction of future technical progress was entrusted to another body which was to be quickly created as a centre to determine the basic lines of technical policy for each branch and for industry as a whole. It was to be attached to the chairman of VSNKh himself, to be called the Supreme Expert Council and to include industrial managers and technical specialists.[63] At the same time the Presidium of VSNKh resolved to include the heads of *glavki* on the board of NTU, in the hope that this would improve the supervision of the NTSy; the official view was that the latter were conservative in their attitude to advanced technology and were proving inadequate as representatives of science and technology in the compilation of the five year plan.[64]

Against this background the results of the survey of the industrial research network appear to have been surprisingly favourable to NTU, with only relatively minor criticisms of its role in organising the research network and of its failure to provide full assistance to the research institutes.[65] Moreover on these matters it is reported to have been defended strongly by VSNKh in its remarks to Sovnarkom on the survey report.[66] In fact, in the preamble to the measures approved by Sovnarkom[67] on the basis of this report, one of the four reasons cited for the 'significant' improvement in research in the immediately preceding period was 'the union of scientific and technical organs under a single administration of VSNKh'. A significant recent improvement in NTU's direction (*rukovodstvo*) of the research institutes was also noted, thereby suggesting that as an administrative organ it was successfully coping with the rapid changes that were taking place in industrial research. Furthermore, the faults in the linking of research with industry were not only attributed to the insufficient connection between the work of the research establishments and the current and long-term needs of industry, but also to a lack of attention by industry and the conservatism of the upper echelons of industrial management. The decree proposed no radical changes in the activity of NTU; it primarily called on VSNKh to generally strengthen the work of NTU.

However, while this decree confirmed NTU's 'science policy' role, it contained no reference to its role in helping to make or review technical policy, which had, of course, already been diminished by the formation of the Supreme Expert Council.[68] The fact that there was nothing on the NTSy was very significant; it suggested that the future of these bodies was still in the melting-pot in spite of the measures taken by the Presidium of VSNKh in May 1928. This indeed proved to be so, for just over a week later the NTSy were transferred to the authority of the *glavki*.[69]

The decree which put this transfer into action stated that it was not to 'lessen the role and importance of NTU'. Some substance seemed to be given to this assertion, not only by the proposals of the Sovnarkom decree, but also by a widening of the activities of NTU, for during 1928 it took control of two further organisations. In May a new body had been set up under it to stimulate Soviet inventions and to take measures to introduce both Soviet and foreign inventions into industry – the Central Bureau for the Realisation of Inventions.[70] In December the transfer of the Geological Committee, the USSR's leading body in geological surveying, from the Chief Administration for Mining and the Fuel Industry to NTU was approved.[71]

At the same time as the activities of NTU were being subjected to this critical scrutiny a completely new body appeared on the scientific scene which presented a potential threat to its position as the body responsible for organising and coordinating research of interest to industry and for linking science, technology and industry. This was the Committee for the Chemicalisation of the USSR which was established under Sovnarkom in April 1928.[72] The initial movement for the creation of such a body was the result of a dissatisfaction with government policy in the field of science on the part of the country's chemists, who were at the time probably the Soviet Union's strongest scientific group. In March 1928 a number of chemists, headed by A. N. Bakh, who had been critical of the work of NTO earlier in the twenties,[73] petitioned the government to establish a special commission to assist the expansion of work in chemistry. They thought that chemistry and the chemical industry were not being given the necessary level of priority in the plans for industrialisation.[74] The new committee was to be chaired by Rudzutak, a deputy chairman of Sovnarkom, his deputies were Krzhizhanovskii – head of Gosplan, Kuibyshev and Bakh; its membership comprised both scientists and the heads of branches of the chemical industry and the directors of important chemical plants.[75] It was intended to be a chemical equivalent of GOELRO and, among other things, to finance research and arrange and aid international scientific contacts. Clearly, if it fulfilled its intended functions, this new committee was going to usurp some of the functions of NTU.

Nevertheless, in spite of this new development and its loss of control of the NTSy, NTU was still a very important body in science and technology. Indeed, at the Fifth Congress of Soviets in May 1929, Pokrovskii, when reviewing the state of the country's R&D network, considered it to be 'more powerful' than the Academy of Sciences.[76] On the other hand what the decree on the NTSy in particular had shown was that VSNKh was now prepared to take administrative measures to link science to industry where it was felt that attempts to strengthen the existing structure or to create new advisory bodies had proved unsuccessful. It made it likely that the research institutes themselves might also at a future date be placed under the control of the *glavki* in an attempt to improve their connection with industry. Furthermore, notwithstanding the general approval of NTU expressed in the decree of August 1928, it would appear that by the start of 1929 it was coming under attack inside VSNKh, notably from Mezhlauk, a former head of the iron and steel industry, who had been appointed a deputy chairman of VSNKh in July 1928. A Soviet author who was working for the VSNKh newspaper at the time writes of Mezhlauk as follows:

> Having acquainted himself with its (NTU's) work, he subjected it to sharp criticism. There was no initiative, no originality. The activity of the institutes was not linked to the work of industry. The institutes, to the greatest possible extent, had to be joined to production, so as to be occupied both on theoretical studies and on making technical experience generally available.[77]

Meanwhile there was now again increasing debate about the industrial administrative structure. A consequence of the increasing investment and planning of the industrialisation programme was the tendency for centralisation also to grow. The new *glavki* established by VSNKh in 1926 were an obvious mechanism by which such an expansion in the central direction might have occurred, especially in the light of their role at the beginning of the 1920s. However, they did not achieve the dominance over industry of the *glavki* of the earlier period. In 1928 complaints of 'glavkism' were being aimed not at the *glavki* themselves but at the syndicates (*sindikaty*). The latter had originally been established by the trusts as jointly owned sales organisations when they were confronted by the market-oriented conditions of the New Economic Policy. Over the succeeding years they had continuously increased the proportion of sales which they controlled and begun directly to influence the size and assortment of production through advance supply contracts with the trusts. By 1928 they were the main organs of detailed planning and control in industry. The planning functions performed by the *glavki* were now appearing largely superfluous and in the textile industry the process had gone so far as to result in the abolition of the *glavk* at the beginning of that year.

At the Sixteenth Party Conference in April 1929 the continuing fear of bureaucratic central control of the economy resulted in heated discussion of a report by Yakovlev of the Central Control Commission on 'the results and next tasks of the struggle with bureaucracy' and the subsequent resolution referred to the need to improve the administration of industry. The conference agreed that the independence of the factory should be increased and that VSNKh should be transformed into an organ 'not only of planning and economic control . . . but also of active technical leadership based on the achievements of both American and European and Soviet science and technology'.[78]

It was during the discussion of the details of the proposed changes that the decisive clash over NTU was to take place. The finally agreed solution to the interlinked problems of eliminating bureaucratic competition among *glavki*, syndicates and trusts and of securing the desired decentralisation in decision-making was to liquidate the *glavki*, drastically cut the powers of the trusts and establish on the basis of the syndicates 'combines' (*ob"edineniya*), to control the individual branches of industry. VSNKh's central apparatus was to have purely an overall planning and coordinating role. The predominant question concerning the research facilities of VSNKh was whether they should be attached to the *ob"edineniya* which were to administer industry and, if so, what the role of NTU was to be in the future. Some people were apparently of the opinion that the institute system should remain independent and serve as the scientific and technical foundation for the work of the Supreme Expert Council which had been established in the preceding year.[79] However, in September 1929 the VSNKh commission dealing with the reorganisation of the industrial administration is reported to have proposed that, while the plans of the scientific and technical institutes should be approved at the centre, the institutes themselves should be, with the NTSy, administratively subordinate to the *ob"edineniya*.[80]

When, shortly afterwards, these proposals were reviewed by the

Presidium of VSNKh, it became clear that matters were not as simple as this earlier report suggested. There were, in fact, two closely interwoven topics under discussion: firstly, the future role of NTU and its existing institutes; secondly, the structure of the envisaged *ob"edineniya* and their provision with scientific services. While there appeared to be a measure of general agreement that the *ob"edineniya* should be provided with strong central scientific laboratories, there was no consensus on whether these should be completely new bodies or whether their functions should be carried out by the existing NTU research institutes. Indeed the editors of the VSNKh newspaper noted that there was a 'significant divergence of opinion ... in connection with the review of the place of NTU and the research institutes in the industrial system'.[81] No less a person than Bukharin fought a strong rearguard action against such a decentralisation of research.[82] Bukharin had been appointed head of NTU earlier in the year,[83] replacing Kamenev who had briefly held the post since the previous September.[84] It was probably intended that Sverdlov would continue to run the organisation, as had Ipatieff under Trotsky, and he may have remained in full control under Kamenev, but Bukharin was to play a very active role and had previously shown himself to be interested in the role of science under socialism.[85] His views on the control of industrial research were in many respects to parallel closely those expressed by Dzerzhinskii in the discussion of the future of NTO four years before. They were summed up in the title to the report of his remarks: 'Laboratories – to the enterprises, institutes – to NTU.' While agreeing that enterprises and *ob"edineniya* should have their own laboratories, he argued against a policy of transforming NTU's research institutes into such laboratories. He considered that any decentralisation of research would only be a short-term solution; the use of scientific manpower in the way implied in the proposals, namely to eliminate the lack of middle-level (*srednyi*) technical personnel, would help in the immediate future, but was in fact the worst way to make use of the existing research manpower. If this policy was pursued and the institutes turned into the central laboratories of the *ob"edineniya*, then at some future date these same institutes would have to be re-created in order to maintain longer-term research work. Bukharin was supported by Shein, who also emphasised that the only way to prevent the future development of industry being jeopardised by the daily scientific needs of industry was by keeping a single central organ to control the research institutes.

However, all the other speakers whose remarks were reported

emphasised the great gap between the institutes and industry. For example, Mezhlauk called NTU a 'contemplative' body and Likachev, the director of the AMO automobile factory, said that his plant got almost no assistance from the Scientific Automobile Engineering Institute. On the other hand there was some divergence of opinion on the degree of decentralisation needed. Yulin, head of Glavkhim, felt that those institutes which were not directly connected with particular branches of industry should be left under NTU; he specifically mentioned the Karpov Chemical Institute and the State Physical Technical Institute. Mezhlauk seems to have implied that every institute should be under an *ob"edinenie* – he wanted no theoretical institutes at all. While he agreed that there had to be departments occupied on fundamental research, he found it impossible to accept that in a huge institute there should be only one small department engaged on work of direct relevance to industry.

Summing up at the Presidium meeting Kuibyshev envisaged the continued existence of NTU but in a changed form. It was being converted from 'an organ in charge only of institutes ... into the headquarters (*shtab*) of the scientific and technical leadership of industry'. Thus it was, in the technical field, to complement a second 'headquarters' which was to be created in the economic planning field by the strengthening of PEU, the Planning and Economic Administration. However, he said, this did not mean that research should be forgotten; NTU would still control the institutes which did work of interest to more than one branch of industry and the more specialised institutes would be jointly controlled by NTU and the *ob"edineniya*. Further, he considered that the central laboratories of the *ob"edineniya* should do research. The import of the last remark became clear in the subsequently published theses of VSNKh on the reform. While they stated that NTU 'directly administers (*upravlyaet*) the whole system of research institutes', in the very same paragraph it said that the network of research institutes should be reviewed 'with the aim of converting individual institutes into the central laboratories of the *ob"edineniya*'; NTU would direct (*rukovodit'*) their work.[86]

Later in September 1929 Gosplan approved the reorganisation proposals of the Presidium of VSNKh.[87] Nevertheless in the following month, at the Sixth Plenum of VSNKh, Bukharin continued to fight for the retention of all research institutes under NTU.[88] But, by this time, the major question concerning the research network would appear to have been changing from whether NTU should retain control of the research institutes to whether NTU itself was to continue in existence.

Uncertainty about the actual nature of the reorganisation of the central apparatus of VSNKh was reflected in the published theses of the Central Control Commission on the projected reform; for, while going into detail on the new administrative structure for each branch of industry, they contained nothing on the future structure of VSNKh itself.[89] In fact, in this respect, the final form of the reorganisation which was approved by the Central Committee on 5 December showed a radical change from VSNKh's original proposals.[90] NTU and PEU were not to exist as separate departments. The resolution stated that 'the strengthening of the technical aspects of the ... planning of industry demands the creation in VSNKh SSSR of a single organ of technical and economic planning'. The functions of this body, PTEU, Planning, Technical and Economic Administration, were thus to be: firstly, the working out of long-range plans and control figures, the planning of the geographical distribution of industry, the coordination of the activities of the *ob"edineniya* and the review of industrial policy; secondly, the control of the technical reconstruction of industry, the development of future technical policy, the overall control of research, the administration of the institutes directly subordinate to it and the control of the transfer of foreign experience.

It seems more than probable that this final amalgamation of NTU and PEU was linked to the expulsion of Bukharin from the Politbureau of the Communist Party at the plenum of its Central Committee which was held in November 1929. His views on NTU are known to have been attacked by Kaganovich[91] and the abolition of NTU as a separate entity, which was to mean that the research network was to come under a sector of a larger body, also served to reduce the position of Bukharin inside VSNKh.

The reorganisation, as approved by the Central Committee's resolution, was implemented, with respect to NTU, by a VSNKh decree of 30 December 1929.[92] Seventeen research and scientific establishments were definitely to remain centrally administered, i.e. under PTEU, two more while a special investigation of them was undertaken.[93] The remainder, eighteen in all, were subordinated to the *ob"edineniya*.[94] The bodies concerned with inventions and NTU's publishing houses were to come under PTEU. Of the other organisations which it controlled, the Geological Committee had been turned into an independent Chief Geological Survey Administration in October 1929;[95] the Chief Inspectorate of Weights and Measures was transferred to the All-Union Committee on Standardisation under STO;[96] NTU's Rationalisation and Standardisation Department was, in February

1930, combined with two organs of PEU into PTEU's Sector for the Rationalisation of Costs and Labour.[97]

When NTU had finished supervising these changes it was itself liquidated. At the same time a research sector, NIS, was established in PTEU to administer the research institutes which it directly controlled. Attached to it was to be a central research council, the Central Scientific Research Council for Industry, with a membership of fifteen. Bukharin was appointed head of NIS and chairman of the council.[98]

5 Decentralisation and Industrial Research, 1930–40

THE ADMINISTRATIVE STRUCTURE FOR INDUSTRIAL RESEARCH UNDER VSNKh AND NKTP, 1930–6

The first years of the 1930s were a period of almost constant change in the industrial administrative structure, and since the majority of VSNKh's research institutes were now directly controlled by industrial organisations, these changes had important repercussions on the research network. The basic outlines of the changes were as follows: the large *ob"edineniya* which had been set up at the end of 1929 proved to be very cumbersome and soon developed the unwieldy administrative bureaucracies which their formation was specifically designed to avoid. Further, notwithstanding the establishment of each *ob"edinenie* as an organisation to exercise full administrative control over a branch of industry, in practice the Presidium of VSNKh itself took many of the major policy decisions relating to the development of particular branches. Consequently this attempt to decentralise the control of industry by forming these organisations outside the central apparatus of the commissariat was to be short-lived. Within a very short time the *ob"edineniya* began to be broken up into smaller units of a more manageable size. As a result some branches of industry (for example, coal and steel) were soon to come under the control of more than one *ob"edinenie* and increased responsibility for the development of the branch as a whole devolved on VSNKh itself. This was an important factor behind a reform of the central apparatus of VSNKh in November 1930 after Ordzhonikidze had taken over as its head. At this time PTEU, which had also in its turn proved to be a very cumbersome body, was broken up to form twelve functional sectors, of which a research sector, NIS, was one, and seven branch sectors under which the *ob"edineniya* were grouped. From the middle of 1931 the transfer of authority from the *ob"edineniya* to the commissariat's central apparatus accelerated as the branch sectors took on some sales and supply functions and began to be reorganised into *glavki*. At the same time the process of breaking down the *ob"edineniya* was also accelerated. Individual *ob"edineniya* were either divided into several smaller *ob"edineniya* or split up on the basis of their constituent trusts.

The move towards a more specialised administrative structure led to a reorganisation of VSNKh itself at the start of 1932. It was divided into three commissariats. The timber and light industries were each now to come under their own commissariat, NKLesProm and NKLegProm respectively. At the same time the remainder of VSNKh was renamed the People's Commissariat for Heavy Industry, NKTP. The republican VSNKhy became republican commissariats for light industry. Meanwhile the division and reorganisation of the *ob"edineniya* continued. By the end of 1932, for example, NKTP's chemical *ob"edinenie*, Vsekhimprom, had been split into fourteen trusts and *ob"edineniya*.[1] In 1933 the emphasis switched to the complete abolition of the *ob"edineniya*, which now retained little of their original functions, and by the end of the year few were still in existence. The *glavki* were now undergoing the same process of size reduction and increased specialisation as had the *ob"edineniya*. The *glavk* for the chemical industry, for example, was split into four.[2] Whereas in mid-1933 there were seventeen branch production *glavki* under NKTP, at the end of the year there were twenty-nine;[3] and two years later there were at least thirty-two.[4] The increasing size and diversification of Soviet industry was thus reflected in a growing specialisation of the administrative system. This was also intended to improve the links between the commissariat and the factory.

This almost continuous reorganisation in the structure of the industrial administration frequently led to changes in the organisational structure of industrial research; and with the additional factor of the rapid foundation and, at the end of 1932, dissolution of research establishments, it is difficult to arrive at a clear picture of all the administrative developments in the research network.

We have seen that, at the end of 1929, only seventeen of the research institutes previously under NTU were put under the direct control of NIS PTEU VSNKh – between a third and a half of the total. A further erosion of the direct control of research facilities by the central VSNKh organ responsible for science occurred at the time of the reform of its central apparatus in November 1930. The discussion of the proposed reform appears to have led to further debate on the organisation of research. On 16 November the VSNKh newspaper, *Za Industrializatsiyu*, carried a series of articles criticising the work of the research establishments and the activity of NIS PTEU.[5] These were accompanied by editorial comment pointing out the failure of the industrial research institutes to fulfil the role allotted to them and blaming NIS PTEU for this situation. It was suggested in one of the articles that all the institutes which it directly administered should be immediately transferred to the *ob"edineniya*, leaving NIS with only a general supervisory function. The decree on the new structure of VSNKh did not go quite as far as this, but from the point of view of NIS the difference was minimal. The decree's provisions envisaged that, in future, only Ioffe's State Physical Technical Institute would be directly under the new NIS. Thirteen research institutes were to be transferred from NIS to VSNKh's new branch sectors and two to functional sectors other than NIS.[6]

The attitude to these changes of Bukharin, who on the same day was appointed head of the new NIS,[7] is unknown. Nevertheless, while no explicit mention of any opposition has been found, it would seem likely that he again took the position of the previous year and strongly supported the retention of the research institutes under NIS. Criticism of the change is implied, perhaps, in a newspaper article by him which was published three weeks later.[8] On the one hand he attacked the attitude taken by the *ob"edineniya* towards science, and on the other hand he suggested a three-tiered structure for the VSNKh research network – which stressed that there was an important role for institutes which were not narrowly specialised. Bukharin's three categories were, in order of decreasing specialisation, factory laboratories, branch (*otraslevye*) institutes and head (*golovnye*) institutes. The head institutes were to provide the theoretical and methodological basis for the work of the branch institutes and factory laboratories, to set out the main directions for future applied research and to provide the basis for the planning of scientific research. Bukharin undoubtedly envisaged that such institutes should be directly under NIS, VSNKh's chief scientific organ. However, as in the discussion of the reorganisation of 1929, wider political issues were also probably involved, for at this time a campaign was being carried out against 'right deviationists', i.e. Bukharin.[9] Indeed, he was personally criticised in the previously cited issue of *Za Industrializatsiyu* for 16 November for a visit to the Mekhanobr institute during which he considered it 'completely unnecessary' to visit the party collective and other social organs. Furthermore, at the end of November 1930 the trial of the 'Industrial Party' began and many of the defendants were specialists who had been engaged on work in close contact with the research network of VSNKh; the so-called leader, Ramzin, had earlier taken quite an active part in the work of NTU VSNKh.[10]

The decree of 29 November marked the beginning of a period of rapid change in the administrative subordination for some VSNKh research institutes. There would appear, however, to be some doubt as to whether all the changes envisaged in this and subsequent decrees were ever fully implemented and in some instances, at least, they were quickly superseded. Thus, some institutes which were originally to come under a VSNKh production sector were soon transferred to the authority of *ob"edineniya*. For example, a VSNKh decree of December 1930 which established the all-union *ob"edinenie* for the optico-mechanical industry stated that it was to control the Technical Division of the State Optical Institute,[11] and just over a fortnight later five other establishments which had come under the sector for the chemical industry were transferred to the *ob"edinenie* for that industry.[12] One of these was the Karpov Chemical Institute; yet a resolution of the Presidium of VSNKh of February 1931 stated that it had decided to keep (*ostavit'*) the Karpov Institute and Ioffe's institute under NIS as head (*golovnye*) institutes.[13] At about the same time a further decree was published concerning the Technical Division of the State Optical Institute; this also stated that it was to be kept directly under NIS, while adding that it was to do everything possible to meet the demands of the factories of the *ob"edinenie* for the optico-mechanical industry.[14] Later, in June 1931, the Mekhanobr institute was transferred from VSNKh's mining and ore sector to NIS.[15] These changes clearly suggest that NIS may have been engaged on a successful rearguard action aimed at maintaining its direct administrative control of some research institutes. In fact, as a result of further changes embodied in a decree of August 1931, NIS was firmly re-established as an organ which was not only responsible for the planning and coordination of research but which also administered part of the research network.[16] This particular decree was the result of a review of VSNKh's research establishments undertaken in connection with the envisaged reorganisation in the structure of the *ob"edineniya*.[17] The breaking down of these into more narrowly specialised units meant that some institutes under an *ob"edinenie* would in future be undertaking research of interest to more than one; these and the institutes under VSNKh's production sectors were now to join the 'theoretical' institutes under the direct control of NIS. This group comprised twenty out of the eighty-three institutes named in the decree, The remaining sixty-three were divided among thirty-seven *ob"edineniya*, the recently established *glavk* for the timber industry and one trust.[18]

Meanwhile the rapid expansion in the research network continued and while the actual details of this growth are not known, it would seem likely that it was closely associated with the rapid increase in the total number of *ob"edineniya* and trusts directly subordinate to VSNKh and NKTP. Each newly established body would undoubtedly have wanted to organise its own research institute, to have 'its science' as Pokrovskii had said at the Sixteenth Party Conference when referring to the desire of commissariats to have their own research institutes.

By mid-1932 a move to check this growth had started within NKTP. On 29 July a decree was issued specifically devoted to the question of the organisation of new research institutes.[19] Its measures reflected a lack of central control over the hectic development of the research network that was in progress. They were aimed, firstly, at avoiding the formation of new institutes which were surplus to requirements and poorly equipped. The opening of an institute without the permission of the commissariat's board was forbidden, and it was stated that the approach for permission could only be made by NIS and that NIS, when presenting the proposal, must produce detailed information on the need for the institute and on the availability of funds, equipment and manpower. Secondly it was proposed that within a month NIS should review the research network and present the board with draft proposals for the abolition of 'feeble' (*nizhnesposobnye*) institutes.

An important contributory factor to this growing dissatisfaction with the rapid growth of these new institutes must have been the financial position of research under NKTP. For while in the previous two years the funds going to research under VSNKh had nearly tripled in real terms, in 1932 real expenditure on all industrial research establishments was in fact going to be only marginally higher than real expenditure under VSNKh alone in 1931.[20] This slackening in the rate of growth of funds was a reflection of the growing economic difficulties but the effect of these difficulties was undoubtedly magnified by a further switch towards indirect finance from industry. In 1931 funds from industry comprised 40 per cent of the expenditure on VSNKh's research establishments, in 1932 80 per cent of the funds for NKTP's research establishments came from industrial organisations.[21] It seems probable that the growing economic strain would have had a considerably greater effect on such indirect finance than on central budget allocations.

It has been impossible to discover whether the NKTP decree of 29

July was part of a general review of the country's entire research network or purely a development internal to NKTP.[22] What is known is that on 26 December 1932 Sovnarkom issued a decree 'On the reorganisation of the network of research institutes of NKTP' which was based on material provided by NKRKI.[23] Five days later NKTP passed a decree to put Sovnarkom's proposals into action; it gave NIS NKTP twenty days to put through the proposed amalgamations and liquidations.[24]

According to the published official data, the number of NKTP's research institutes was cut at this time from 162 to 136.[25] This reorganisation meant that many of the institutes established in 1931 and 1932 had a very short life. A typical example is provided by the development of the research facilities servicing the fields of construction and building materials. In November 1931 VSNKh had twelve institutes (with twenty-three branches) working in these fields.[26] The *Za Industrializatsiyu* article of 22 December 1932 speaks of the recent setting up of sixty research establishments and proceeds to criticise sharply the resulting situation: the majority did no original work; in several cases there was a large overlap between institutes which were in the same region of the country; several small institutes (*institutki*) had been established to work in narrowly specialised fields which were more than amply covered by large institutes such as the Central Research Institute for Building Materials in Moscow, which did research in all areas relevant to the industry. The reorganisation swept away most of these establishments. In April 1933 twenty-two institutes (with five branches) were listed as working on the problems connected with building and building materials;[27] at least seventeen of these had been in existence in 1931, ten as institutes and seven as branches of institutes.[28]

While the reorganisation at the beginning of 1933 envisaged no radical change in the administrative structure for industrial research, there was continuing discussion of the links between research institutes and industry. A resolution of the Second All-Union Conference for the Planning of Research in Heavy Industry, held in December 1932, considered it necessary to 'attach' (*prikrepit'*) individual research institutes to particular new factories.[29] In an article on the conference A. A. Armand, Bukharin's deputy at NIS, said such a move was needed because 'the subordination of institutes to the *ob"edineniya* and *glavki* has failed and is still failing in practice to provide the necessary links between the work of scientific establishments and enterprises.'[30] 'Attachment' of institutes to plants was envisaged by the resolution of NKTP's board which reviewed the work of the planning conference.[31]

Pressure for a related change in the actual research administrative structure appears to have built up later in 1933. Discussion was probably initiated in connection with the new policy of abolishing the *ob"edineniya*. An editorial in an issue of NIS NKTP's house journal published in September 1933 dealt with the research network under the heading 'The reconstruction of administration and the scientific servicing of industry'.[32] After criticising the work of the specialised industrial research institutes, it suggested that research establishments should be transferred to the country's leading factories to create 'a powerful scientific and technical fist, a centre of technical culture in the enterprises'. The revised structure for research would still have institutes doing fundamental work directly under the commissariat; others would provide services for the *glavki*; the rest would go to the plants.

However, despite this statement in an authoritative source, no such radical administrative reform was undertaken at this time, and throughout the first half of the 1930s the changes in the administration of research were to reflect closely the changes in the administration of industry at the branch level. Institutes under *ob"edineniya* in 1931 were by the beginning of 1934 frequently under the trusts which had been established in the intervening years. Such a development was especially true of the chemical industry. As of the breakdown of the *ob"edineniya* there were eight in various branches of that industry with fifteen institutes under them.[33] In 1934 there were four *glavki* controlling the industry. These directly administered four institutes and eleven of the trusts under these *glavki* controlled twelve institutes.[34] In 1931 the institutes for chemical dyestuffs, plastics, pharmaceuticals and paints were under the *ob"edineniya* responsible for these branches of the chemical industry: by 1934 these branches and their institutes were under trusts of Glavkhimprom, the *glavk* for inorganic chemicals. In NKTP as a whole in 1935 thirty-seven institutes were directly under *glavki*, twenty-eight under trusts and two under one of the few remaining *ob"edineniya*; since these institutes were under a total of twenty-nine *glavki* and twenty-five trusts, it can be seen that industrial organisations usually controlled only a single institute.[35] Under the trusts there were, in the main, two types of institute. In the first place some trusts were responsible for a whole sub-branch of an industry and for these it was natural that the research institute concerned with this field should be attached to the trust. In the chemical industry, the institutes for dyestuffs, plastics, pharmaceuticals and paints are obvious examples. Secondly, some trusts established on a geographical basis had research institutes attached to them; for example, the Groznyi Oil Trust, the Eastern Steel Trust (Vostokstal'), the Urals Trust for Non-Ferrous Metals and the Ukraine Chemical Trust all had institutes. This second development was part of the policy of expanding research in the provinces.

Dissatisfaction with the work of the industrial research network was, however, still leading to a questioning of the ability of this administrative structure to produce the desired linkages between science and industry. In 1936 another review of the work of the industrial research institutes was set in motion and it became clear that as in 1933 there was a movement to undertake a radical reorganisation of research and to transfer research institutes to factories. It is, indeed, reported that the primary aim of the review was to ensure the maximum development of scientific and technical work at factories by transferring research and design resources to factory laboratories and design bureaux. At the same time there were worries about the low quality of some research personnel and NIS was to review the staffs of research establishments and fire low-quality personnel. A new departure was the proposal that some poorly staffed institutes should be absorbed by higher educational establishments; this was a clear reflection of the new and growing emphasis on the research activities of the higher education sector.[36] Judging from the discussion of the state of research and of these proposals which subsequently took place, the abolition of NIS itself was also being suggested, perhaps as a corollary of measures which were being taken to strengthen the role of the *glavki*. Not surprisingly, at a meeting on the reorganisation which was called in August, the commissariat's leading scientists defended the work of their institutes and also the role of NIS. In their turn, they criticised the *glavki* for not providing the necessary level of direction for the work of the institutes. Although there was agreement that some institutes could be placed under the control of the country's leading plants, the idea that there should be a blanket decentralisation of this type was strongly opposed. It was further stressed that the 'academic' institutes, as one scientist described institutes such as the Karpov Chemical Institute and Ioffe's physics institute, should be under a special department of science. But, on this last point implicit dissent to the maintenance of a central core of institutes doing fundamental work was expressed by Academician Bardin, who considered that institutes and laboratories doing 'long-term' work should be in the Academy of Sciences.[37]

The outcome of the review of industrial research was that no blanket decision was taken on the structure of NKTP's research network. Bauman, head of the newly established Central Committee department of science, himself pointed out at the meeting in the commissariat in August that it was not possible to provide one all-embracing answer to the question of the correct way to organise the network of research institutes.[38] Nevertheless, the changes which were implemented were as far-reaching as the reorganisation at the end of 1932. From published reports on the progress of the review it would seem that six or seven institutes and branch institutes were closed down completely and that similar numbers of institutes were both transferred to factories and placed under the control of particular trusts. A smaller number were amalgamated with NKTP's technical institutes.[39] At the same time some institutes were transferred from the authority of NIS to the control of *glavki*. In all around fifty institutes were affected to a greater or lesser extent by the reorganisation. Thus, while there was not the comprehensive linking of research establishments to factories as appears to have been originally intended when the review was set in motion, the discussions of 1936, unlike those of 1933, did lead to the devolution of some of NKTP's R&D resources to the enterprises. Yet the changes in the administrative structure which took place basically continued the process of attaching the institutes to *glavki* and trusts which had been taking place over the previous five years, and NIS resisted whatever attempt was made to abolish it. The reorganisation's greatest effect was not on administration but on the size of the industrial research network.

The administrative decentralisation of industrial research which had been taking place in the first half of the 1930s was matched by a parallel decentralisation in the funding of the industrial research establishments. The sharp increase in the indirect finance of research which had taken place in 1932 had been maintained; in 1934 only 15 per cent of research funds came directly from the state budget.[40]

INDUSTRIAL RESEARCH IN OTHER COMMISSARIATS, 1930–6

During the first half of the 1930s the number of commissariats responsible for various parts of industry grew. The move away from a position where the whole of industry was run by a single commissariat began in 1930 when the control of the food industry was moved from

VSNKh to Narkomtorg, the commissariat for trade[41] – from the end of that year it came under NKSnab, the commissariat for supply,[42] and in 1934 a commissariat (NKPP) solely for the food industry was established.[43] Then in 1932 light industry and the timber industry became the responsibility of separate commissariats. All three industries had their own research institutes but their research networks were far smaller than that of NKTP – at the beginning of April 1935 their total manpower was under ten thousand, which was less than a third of the number employed in NKTP's institutes.[44] Available information suggests that only NKLegProm and NKSnab had a sector of their central apparatus specifically devoted to matters of research; and when as a result of a 1934 reorganisation of NKLegProm the central administrative structure was greatly reduced, the functions of its Sector of Research Establishments were handed over to its *glavki*.[45]

A further small number of research institutes had been under the republican VSNKhy, which were converted into republican commissariats for light industry at the end of 1932.[46] Some institutes which specialised in other fields than light industry were transferred to NKTP.[47] Just over two years later, in August 1934, the republican commissariats for light industry became the basis for new commissariats for local industry. Thus, for example, the Research Institute for the Musical Industry was under VSNKh RSFSR at the end of 1931, NKLegProm RSFSR in 1932 and NKMestProm RSFSR in October 1934.[48] In March 1936 just over 2000 were employed in research establishments under NKMestPromy, the commissariats of local industry.[49]

THE INDUSTRIAL RESEARCH ESTABLISHMENTS AND THE GROWTH OF COMMISSARIATS, 1937–40

Although by the mid-1930s there were four all-union commissariats and also republican commissariats responsible for industry, NKTP clearly had a predominant position, controlling as it did all the priority sectors of Soviet industry. Similarly the major part of the Soviet industrial R&D network was controlled by this one commissariat. On the eve of the Soviet entry into the Second World War the administrative structure of Soviet industry was to be greatly changed; these changes had similarly radical consequences for the organisation of industrial research. In their broad outlines these developments reflected a continuance of the increasing administrative specialisation which had taken place within NKTP during the first half of the 1930s as the industrial base had expanded and new branches of industry had been established. The important difference was that they took what was the obvious next step, the formation of more specialised commissariats in the priority areas of industry. In December 1936 a special commissariat was set up for the defence industries, which had been previously under NKTP, although subject to a greater degree of control by the leadership than the rest of industry. Next, in August 1937 NKTP's engineering *glavki* were separated off to form an independent commissariat of engineering. At the same time some of the very large *glavki* in NKTP were split into smaller more specialised *glavki* – GUMP, the giant *glavk* for metallurgy, was divided into five.[50] Finally, in the period from the beginning of 1939 to the German invasion, a whole new series of commissariats were set up. At the beginning of 1939 the Commissariat of the Defence Industry was split into four, the Commissariat of Engineering into three. NKTP itself became six separate commissariats and NKLegProm and NKPP were divided into two and three commissariats respectively. By July 1941 this process had been taken even further and Soviet industry on the national level was administered by twenty-four commissariats and there were other republican commissariats for the local consumer goods industries.

The formation of separate commissariats for the defence industries and engineering roughly halved the size of NKTP and meant that it lost about twenty of its research institutes, including Ioffe's Leningrad Physical Technical Institute, which was now to come under NK-Mashinostroenie, the commissariat for engineering. Furthermore in the autumn of 1937 the powers of the *glavki* were again increased. These two administrative developments were to lead to the final demise of NIS NKTP and the end of a core of centrally controlled research institutes; for when in November 1937 a new statute for NKTP was approved to take account of these changes, it stated that the direct control of the industrial research institutes was vested in the technical departments which all *glavki* were to have, with the general supervision of the research institutes becoming a function of its Technical Council, the body responsible for reviewing the commissariat's technical policy.[51] This structure was also shared by other commissariats.[52]

Pressure also appears to have continued for the closure of institutes as independent organisations and the use of their staff and resources to strengthen research at enterprises and the technical institutes which trained industrial specialists.[53] Employment in NKTP research institutes does seem to have fallen in 1938 and at the beginning of 1939 probably only just over a half of the personnel in industrial research institutes were working under NKTP.[54]

The break-up of NKTP into a series of commissariats at the start of 1939 triggered a further discussion of the organisation of research and apparently some further pruning of research staffs.[55] The new commissariats were very narrowly specialised, in many cases covering branches of industry which had been under a single *glavk* in the mid-1930s; their constituent *glavki* correspondingly covered very narrow areas of industry. It was probably this narrow specialism which led to a call for the process of decentralisation to be reversed and for the research institutes, and also the project organisations, to be controlled again by the central apparatus of the commissariat instead of being distributed among the *glavki*.[56] However, there are no reports of any such change being introduced.

Thus, during the 1930s the decentralisation of research, which had begun at the end of the 1920s, accelerated and the central organ which had been responsible for science and technology under VSNKh, and which survived the debates at the end of 1929 only with diminished authority, was finally abolished in 1937 at a time when the major commissariat for industry of which it was a part was itself partly broken up. As we have seen, on the eve of the Soviet Union's entry into the Second World War there were twenty-four commissariats each with its own R&D network. Within each commissariat the research institutes were subordinate to the individual *glavki* and the supervision of their work was the responsibility of the technical council whose primary function was the review of the industry's technical policy.

6 The Central Coordination of Industrial Research, 1930–40

NIS AND THE COORDINATION OF INDUSTRIAL RESEARCH

A corollary of the changes in the administrative system was that the coordination and general supervision of the research establishments became a matter of increasing importance. With the first steps towards decentralisation in 1929, this became the role of NIS PTEU VSNKh.[1] Its functions were spelt out at some length in its statute.[2] Firstly, it was to plan the work of the research network, compiling annual and five year plans for research and drawing up general guidelines for the funding of research and manpower training. Secondly, it was to coordinate research by allocating work on complex problems among the various research institutes and reconciling the research being done both within VSNKh and in other commissariats which did work of relevance to industry. Thirdly, it was generally to supervise the research network, ensuring a high standard in the work of the institutes which were now to be administered at branch level. Fourthly, it was responsible for work on the theoretical problems arising from 'the main tasks of the socialist reconstruction of industry' – to be done by the institutes which it still directly controlled. In one respect only did there remain any of NTU's wider concern of linking science and industry; this was a short sub-paragraph which foresaw NIS providing PTEU with plans and materials on the introduction of the achievements of the research institutes into industry and subsequently its supervision of the scientific aspects of the innovation process. Except for this, NIS's functions were to be purely concerned with scientific research. The Central Scientific Research Council for Industry was to provide the specialist basis for its work. It would decide on the basic directions for future research and provide scientific consultative and review services for the Presidium of VSNKh. The council was to be formed by bringing together within one body the scientific and research aspects of the work of the various NTSy. In several respects the role given to NIS PTEU in matters of industrial research reflects that given on a countrywide scale to the research coordinating body established in the Soviet Union at the beginning of the 1960s,[3] although one important difference was that the State Committee for Science and Technology, unlike NIS, was given, in the 1960s, an important role in the import of technology. Another vital difference between NIS and the State Committee for Science and Technology was their status in the hierarchy of government. The status of NIS was lower in the industrial administration than NTU had been. Until Ordzhonikidze's reorganisation NIS's access to the presidium was through PTEU; and even after November 1930 NIS, as one of twelve functional sectors, did not have the importance of NTU, which was one of only two functional administrations in VSNKh at the end of the 1920s. The State Committee for Science and Technology was set up as a committee of the Council of Ministers, the post-war successor of Sovnarkom. Furthermore, at the start of the 1930s, the Committee for Chemicalisation, as a committee attached to Sovnarkom itself, existed as a high-level organ with responsibilities which impinged on those of the central body for coordinating industrial research. By 1930 this committee was sponsoring a growing amount of research, and discussion had begun within its Scientific Commission (headed by Gorbunov), which was responsible for the finance and coordination of research, on the possibility of creating a number of central chemical laboratories.[4] Indeed the proposal for establishing research facilities was soon to be recast as a call for the formation of an academy for the chemical sciences, which would comprise not only the proposed new institutes but also include facilities transferred from other organisations, among them VSNKh.[5] In August 1930 Sovnarkom set aside funds from its reserves to start work on such an academy.[6] However, this marked the zenith of the Committee for Chemicalisation as a body with pretensions to a role in science policy, for its position as a body outside the existing administrative and planning hierarchy was coming under attack;[7] no chemical academy ever was to be built and in February 1931 the committee itself was absorbed into Gosplan.[8]

While this potential threat to the general role of NIS as the coordinator of industrial research did not materialise, there is considerable doubt as to the extent it either did or was able to do the functions outlined in its statute. We have already seen that at the time of the reorganisation of the central apparatus of VSNKh in November 1930 there were articles in an issue of the VSNKh newspaper[9] criticising the current state of the industrial research network. In fact, NIS was singled out for especially sharp attack. The editorial remarks accompanying the articles picked out the lack of direction from NIS as the primary reason for the position of industrial research. This was taken up several times in the articles themselves. While such comment was, as previously suggested, strongly politically motivated – the editorial remarks were entitled 'Facts about the opportunist practice of NIS and its institutes' – it would seem probable that they did also reflect the actual situation in industrial research. In addition to its relatively unimportant position in the hierarchy, NIS found its authority further diminished by having at its head a man under a political cloud. It was also more than likely that at that time of feverish industrial activity research was assigned secondary importance within the enlarged PTEU; that, with the continuing changes in the industrial structure many branch institutes were being neglected by the bodies which were officially supposed to be administering them.[10]

NIS was not, in fact, given the formal authority necessary to carry out the duties outlined in its statute. For example, although it was to be responsible for planning the expansion of the network of industrial research institutes and laboratories, it was not given the power to control fully the establishment of new research organisations. In the VSNKh decree which laid down procedure[11] trusts wishing to set up laboratories needed only to get the permission of their *ob"edinenie*. *Ob"edineniya* wishing to establish institutes or laboratories took the initial decision themselves, but then had to get the approval of the Presidium of VSNKh with NIS acting as the intermediary. NIS seems not to have been able to organise branch institutes on its own initiative.

Both industry's neglect of its research institutes and NIS's lack of control over the development of the research network were discussed at a January 1931 meeting of the VSNKh Presidium on the state of research.[12] The basic document for discussion was a report by Bukharin and the proposals he made were generally accepted. He was, however, criticised for not producing sufficiently concrete proposals for improving the links between industrial organisations and the research institutes, and a special committee was appointed to produce a draft resolution incorporating additional details reflecting this need. Thus, a resolution of the meeting of 23 January was not finally approved until 9 February.[13] This resolution, which dealt mainly with the role of NIS itself, provided evidence of the problems confronting NIS and of the current administrative state of the research network; it suggested that NIS didn't even have the amount of influence over the development of research that had been envisaged in the decree of the previous June. VSNKh considered that four measures were necessary to ensure that NIS could carry out its tasks. The first two were: that the *ob"edineniya* should 'immediately provide full administrative services for the branch institutes and factory laboratories' and that 'the organisation of new institutes and the reorganisation of existing institutes must take place with the knowledge and approval of NIS'. VSNKh's other two proposals were also aimed at increasing the latter's authority. NIS was to take over the planning of the industrial research establishments' capital expenditure (presumably up to that time the responsibility of the *ob"edineniya*) as well as of their current expenditure. Finally all VSNKh's sectors were to discuss all questions related to research with NIS before raising them with the Presidium.

The resolution represented a major statement on NIS's role in coordinating industrial research. Its stress was on the need for closer control over the research done in the institutes, and on ensuring that research was developed in the right areas. In its planning of research NIS was to direct attention to three general areas: to the basic and decisive questions of industrial reconstruction and the main scientific and technical problems which arose from them; to problems directly related to the practical demands of the present; and to the exploitation of the results of up-to-date research abroad. At the same time certain specific fields were picked out for NIS's particular attention in the current year. It was to ensure the completion of research programmes in metallurgy, chemistry, fuel, agricultural engineering and defence; that priority was given to work on quality improvement (especially in metallurgy); that sufficient work was done on freeing the economy from the need to import and towards the saving of fuel, raw materials and metals, and also for the new giant industrial complexes. NIS was to undertake the direct supervision of the most important projects in these areas and provide the presidium with regular progress reports. As a result of NIS being entrusted with these functions it even more resembled the central coordinating body of the 1960s and 1970s.

To ensure that this extensive research programme could be carried out, NIS was assigned certain priority fields of action. Firstly, stress was to be put on the expansion of facilities for research in important branches of industry – metals, coal and electricity supply were mentioned – and the strengthening of regional developments. Secondly, the work of the existing research establishments was to be improved; for example, NIS was to see that the research plans of the branch institutes were based on the relevant branch's economic plan. Thirdly, to strengthen future overall control of the research facilities in each branch of industry, NIS was to take measures to create scientific and technical departments within the *ob"edineniya*; their primary role was to be to ensure swift innovation of the developments produced by institutes and laboratories.

Whether NIS had the authority or the capability to push through this comprehensive programme for developing research is not clear.[14] Subsequently, several new institutes were founded in the provinces for the coal and metallurgical industries.[15] But the proliferation of poorly equipped and staffed institutes, which necessitated the reorganisation at the end of 1932, certainly does not suggest the close control over the development of research which was envisaged both in the original statute for NIS PTEU and in the resolution of February 1931. Indeed an article on NIS's control figures for 1932 states that there would be 129 research institutes under VSNKh at the end of the year;[16] there were actually 162 under NKTP alone.[17] The difference could of course be the result of a later change in the control figures implemented by NIS, but other evidence suggests it was at least partly a consequence of the continuing inability of NIS to exercise any substantial control over the formation of research institutes. The statement made in the VSNKh resolution of February 1931 that NIS had to sanction new institutes, was to be reiterated in a decree of the following July specifically concerning this one matter;[18] it was again to be made in the NKTP decree of July 1932 which dealt with the organisation of industrial research.[19] This latter decree even included a clause stating that it was forbidden to open a research institute without the permission of the board of NKTP, which suggested that industrial organisations were not only failing to consult NIS but even failing to get the official approval of the commissariat itself. It seems clear, therefore, that the rapid growth of the industrial network between the beginning of 1930 and the end of 1932, which was the most distinctive feature of the developments in industrial R&D during the First Five Year Plan, occurred to a large extent without any close direction by NIS VSNKh/NKTP, the central body responsible for industrial science policy. This fact must also have had repercussions on NIS's ability to perform its role in financial and manpower planning, in research coordination and in directing the activities of the research network into the desired fields. It is not inconceivable that its only contact with some institutes was through its role in winding them up.

By 1933 the 'battle' over research planning, of which the discussion and criticism of NIS's research coordinating and planning work was a part, was largely fought and won. The changing economic situation was also now presenting new problems and new policies. The keynote of the Second Five Year Plan was qualitative improvement. It was to be a time when the fullest use was to be made of the base constructed during the first plan. The main slogan was now 'assimilation' (*osvoenie*). The resolution on the results of the first plan and on the annual plan for 1933, which was approved by the joint plenum of the Central Committee and the Central Control Commission of January 1933, stated that in the Second Five Year Plan 'the slogan of the assimilation of new enterprises and new technology' should be added to the existing 'slogan of new construction'.[20] This new emphasis was reflected in the policy towards NIS and the industrial research network; it can be seen in the retrenchment which followed the rationalisation of research facilities at the end of 1932. The need to make the fullest use of the work of the institutes and to get the fullest return for the expenditure on science was now pushed to the fore as the main priority. Consequently the links between industry and the research institutes were again to come under scrutiny. The resolution of the board of NKTP which was based on a consideration of the results of the Second All-Union Conference for the Planning of Research in Heavy Industry specifically dealt with the utilisation of the work of the research institutes.[21] It named some thirty examples of already existing new developments in products and processes which should be immediately introduced and set out guidelines for the utilisation of the results of future research. In both cases NIS was to be responsible for supervising the process.

Another step in the process of increasing NIS's involvement in the application of scientific results in industry was its amalgamation with NKTP's Sector for Technical Propaganda at the end of 1934.[22] The latter was originally established (with Bukharin also as its head) in May 1931.[23] It was engaged in raising the overall level of technical knowledge and it controlled technical publishing houses, financed factory newspapers and organised meetings and discussions. The amalgamation of these two sectors was a natural development of earlier remarks on the need for them to maintain close links.[24]

In spite of the doubts apparently expressed in some quarters, during the review of NKTP's research network in 1936, about the need for a central body such as NIS, the decree which embodied the decisions taken by the commissariat to improve industrial research contained no criticism of its activities. It was in fact the *glavki* who were singled out for failing to provide the necessary direction of the work of the research institutes. Nevertheless the decree did mark a reversal in the previous years' growing emphasis on the role of NIS in innovation, for it simply stated its functions as a body planning and coordinating research, which very closely resembled its role as foreseen on its foundation in 1930.[25] There was no reference to any supervision of the innovation process. However, NIS was to take on the responsibility, through a new department of inventions, for matters concerning the inventions made by workers in plants and organisations under NKTP – a function which had been part of the activity of NTU VSNKh at the end of the 1920s.[26] But, in contrast, at about this time the responsibility for technical propaganda was removed from NIS and vested in Glavvtuz, the commissariat's *glavk* which controlled its higher educational institutes.[27]

However, such changes were to be of little long-term importance, for this was the last reprieve for NIS and in the following year it was to be abolished; thus the central department for science disappeared fourteen years after Pyatakov had first called for its liquidation.

Over the years of its existence this part of the central apparatus of VSNKh and then NKTP had changed significantly. In the years before 1926 NTO, particularly under Ipatieff, had basically been an organisation run by scientists for scientists, providing the research establishments with scientific support services – for example, publishing facilities and equipment manufacture. The provision of scientific advisory services through the NTSy would appear to have played only a subsidiary role in its activities. Subsequently the developing policy for a planned rapid industrial expansion, which increasingly confronted VSNKh with questions of future technical policy, greatly increased the importance of the NTSy and led to a growing stress on the linking of science and industry in the work of NTU. Dissatisfaction with its performance had resulted in the decentralisation of the NTSy. The continuing emphasis on the closer integration of science and industry was reflected in the functions finally entrusted to NIS VSNKh and NIS NKTP after the period of rapid change which started at the end of 1929; for from the early 1930s NIS was seen as playing an important role in supervising the industrial application of scientific ideas and new technical developments. These functions had not been a part of the work of NTO VSNKh at the start of the period. However, the new NIS did not occupy the authoritative position which the

74 *Science and Industrialisation in the USSR*

originally proposed enhanced NTU would have done and subsequent events showed that it could not exercise any degree of strong central direction on the development of industrial research in the following hectic years, nor was it to be successful in ensuring the widespread innovation of the results of Soviet research.

THE ACADEMY OF SCIENCES AND INDUSTRIAL RESEARCH

It can be seen from our preceding discussion of NIS VSNKh/NKTP that as a result of the erosion of the position of the industrial commissariat's department of science at the end of 1929, there arose the beginnings of a vacuum in science policy-making concerned with industrial research and the technical sciences. Later, with the division of the responsibility for the running of industry between a growing number of specialised commissariats, the need for a high-level body to oversee the industrial R&D effort was to increase. As we have seen, for a short period at the very start of the 1930s it looked as though such a role with respect to chemistry and the chemical industry would be fulfilled by the Committee for Chemicalisation. In the long term, however, it was to be the Academy of Sciences of the USSR which was to fulfil this role until its reorganisation at the beginning of the 1960s, which was accompanied by the formation of the State Committee for the Coordination of Scientific Research, the predecessor of the present State Committee for Science and Technology.

Throughout the 1920s there was to be a growing tension between the Academy of Sciences and the government.[28] A new statute was enacted in 1927 to reflect its designation in 1925 as the Academy of Sciences of the USSR and to replace the existing statute of 1836. This marked the beginning of a change in the Academy's orientation by referring to the 'practical application of science'. It also incorporated the expansion of the numbers of academicians from forty-five to seventy (raised again to eighty-five in 1928); this was to be a first step towards the building of an Academy more sympathetic to the aims of the Soviet government. At this time, however, academicians with interests in the technical sciences were playing a more active role in the work of NTU VSNKh than in the Academy – Ipatieff is an obvious example. Further, four academicians were among those chemists who pressed for the new Committee for Chemicalisation which was soon to throw up the idea for a competing academy specifically for the chemical sciences.

The election of new academicians to fill the vacancies created in 1927 and 1928 took place at the beginning of 1929; they included the first four academicians for the technical sciences. The remainder of that year and the start of 1930 saw sharp conflict and reorganisation within the Academy, with pressure now being applied not only from without but also from within by the newly-elected academicians who had been sponsored by the government.[29] The changes were embodied in a further new statute, approved in March 1930, which foresaw an enhanced role for the Academy in Soviet science and which further stressed its role in the applied sciences by implication when referring to its task of directing 'scientific knowledge towards satisfying the needs of the socialist reconstruction of the country'.[30]

In 1931 the first technical research institute was established. A year later fourteen more technical academicians were elected and the first important step was taken in the development of the Academy's role in this field with the formation of the Technical Group in the division of the Academy for mathematics and the natural sciences. As part of its activities it was to look at new areas of technology of potential industrial importance, such as the underground gasification of coal, and to consider particular large-scale problems which needed an interdisciplinary solution.[31] The Academy of Sciences was thus starting to play a more important role in the technical sciences, while the central science department of VSNKh had lost a lot of its authority, and was, in any case, probably too busy coping with the problems resulting from the massive growth of new research establishments to devote itself to the wider questions of the long-term development of science and technology which were within the remit given to it by its statute.

However, as we have noted in our review of the growth of industrial research, the growing role of the Academy in the technical sciences was not to be accompanied by an expansion of its 'in-house' industrial R&D facilities. Its work was to be carried out through commissions devoted to particular problems. These commissions were to cover both long-term projects such as the gasification of coal[32] and also matters of more immediate interest such as the utilisation of slag.[33] Their functions were to review the major questions in their area, compile and propose a list of relevant research projects, coordinate the work of research organisations in the field, provide information and consultative services to these organisations and disseminate research results.[34] The rationale behind such an arrangement was discussed by Volgin, the permanent secretary of the Academy, in his report at its general assembly early in 1934 on its work in 1933.[35] He pointed out that the

Academy's main concern in the technical field was consultative and that since in any case there was a highly developed network of research establishments under the commissariats, particularly NKTP, it was hardly advantageous to create a parallel set of establishments within the Academy. The close links which ought to be formed between the Technical Group's commissions and departmental institutes would mean that when needed the latter's facilities could be utilised by the Academy.[36]

This emphasis on the use of outside facilities did not prevent the establishment of small research bodies within the Academy – particularly from the second half of 1933. Indeed not all the representatives of the technical sciences appear to have been satisfied with the emphasis on research outside the Academy and there would seem to have been some pressure for more substantial provision for research within the Academy. This view was expressed by some of the authors who contributed to a series of articles published in 1934 which discussed the transfer of the Academy from Leningrad to Moscow.[37] At this time A. I. Nekrasov, head of TsAGI, suggested that the Academy should be provided with a well-equipped technical institute.[38] The formation of such a body was subsequently proposed at a meeting of the technical group at the session of the Academy in December 1934 and it was decided that a special institute should be organised for the group's academicians.[39] However, such an institute was not established as a result of these discussions and there was in the short term to be no substantial development of research institutes for the technical sciences.

At the same session of the Academy at which the formation of this institute was discussed the details for the formation of a technical council were approved. This council was to be attached to the General Assembly of the Academy; its chairman was to be Krzhizhanovskii. Its functions were to correspond on a more general level to those envisaged for the commissions which had previously been set up to deal with specific major problems. It was to be responsible for selecting the research problems which were of greatest importance for the economy and for the planning and coordination of work on them both within the Academy and in the departmental research organisations; it was to work on new applications of already existing scientific theories and research results, and to provide technical consultative services to government organs and the commissariats.[40]

Thus, the role ascribed to the Technical Council of the Academy closely reflected that which had previously been entrusted to the central scientific department of VSNKh since its inception in 1918. Further, in the following year the Academy's third new statute of the Soviet period stressed more strongly than ever its role in the applied sciences and created, although not apparently without some opposition,[41] a division specifically for the technical sciences, alongside the existing two divisions, for mathematics and the natural sciences and for the social sciences (and humanities).

Thus by the mid-1930s, as a result of the new enlarged role for the Academy in national science policy-making, which was a consequence of the pressures applied to the Academy and the reorganisation measures implemented over the previous years, and as an outcome of the vacuum in science policy-making for applied research of industrial importance, which resulted from the continuing process of reorganisation in the industrial administration, the Academy of Sciences had become the organisation responsible for directing and supervising applied research of relevance to industry.

The early history of the Soviet Union's work on radiolocation clearly illustrates the growing role of the Academy even in the first half of the 1930s in an advisory and consultative capacity on important projects.[42] While in the 1920s the military would have approached NTO or NTU VSNKh for scientific advice, in 1933 it was to Academy rather than NIS NKTP that they turned during the early stages of work on radiolocation. It was the Academy which, in January 1934, called a conference to discuss the problem, in spite of the fact that scientists working on radiolocation came from institutes such as Ioffe's Physical Technical Institute, the Leningrad Electrophysics Institute, the Ukrainian Physical Technical Institute and the Institute of Telemechanics, which were all subordinate to NIS NKTP, and from the All-Union Electrical Engineering Institute and the Central Radio Laboratory, which were controlled by two of NKTP's *glavki*. On the other hand, the provision of advice and consultation was only a part of the role set out for the Academy in 1935; and in the field of radiolocation the Academy did not apparently coordinate the work of the various organisations involved, nor did it ensure that no duplication of work took place or that there was the necessary liaison among the research teams. Indeed it seems unlikely that the Academy of Sciences played a role in directing technical and industrial research as comprehensive as that laid out in 1935 in the years before the outbreak of the Second World War. The work of the technical division was initially slow to develop,[43] and later in the 1930s the Academy was to come under strong criticism for its lack of attention to the complex problems which were important for the country's development and for showing insufficient appreciation of its leading role in science;[44] its plan of work for 1938 was initially rejected by Sovnarkom on these grounds, and its plans for subsequent years were subjected to considerable scrutiny.[45] Nevertheless by the beginning of the 1940s the Academy was beginning to initiate some complex research programmes involving a wide variety of research establishments. An example of note resulted from the developing interest of Soviet scientists in nuclear research, which led in 1940 to the establishment of a special commission on the problem of uranium. This was to run a research programme which would cover everything from the search for uranium deposits to the production and study of isotopes; it would appear to have included some degree of financial control over the project.[46] However, it was to be the Second World War which led to the Academy's wide involvement in supervising and coordinating technical research programmes.[47]

A further body which may have been involved in coordinating research but about whose activities little is known was the department of science of the Central Committee of the Communist Party, which was established in 1936 under Bauman. However, the available material strongly suggests that no body was responsible for systematically directing the industrial R&D effort of the Soviet Union during the 1930s and that in many cases when particular problems arose or the leadership desired scientific advice it turned to individual scientists and set up *ad hoc* groups.[48]

7 Research Planning

The application of planning to scientific research was probably the feature of the organisation of science in the Soviet Union in the inter-war years which attracted greatest interest elsewhere. The Soviet case was to be frequently cited in the discussion and controversy that took place in Britain in the 1930s on the possibility of directing the power of science to improve the lives of all members of society, which culminated in the publication of J. D. Bernal's *The Social Function of Science*.[1]

The emphasis on the need to plan scientific development in the Soviet Union was a corollary of the growing attention to economic planning. In view of the importance attached to the close link between science and technology and economic development, it was not surprising that the discussion of economic planning was to be accompanied by debate on the possibility of also planning science and technology. Thus, we find Bukharin writing in 1927 that 'planned scientific work ... is the new principle which is quite fundamentally linked to socialism's planned economy.'[2]

The idea of planning was not wholly a phenomenon linked to the industrialisation debates of the mid-1920s. One field in which the need for some form of planning had already been largely accepted by this time was the surveying and study of the country's natural resources. The Academy's commission, KEPS, had indeed started coordinating such work even before the revolution and Gosplan was involved soon after its formation. In 1923 it had organised an All-Russian conference on the study of the country's natural resources and after the conference a bureau was established to organise future congresses.[3] Similarly, by the mid-1920s the research being done within VSNKh for the various branches of industry was discussed and reviewed by conferences called by NTO and the NTSy.[4] There was also an annual meeting of the directors of VSNKh research institutes at which they reported on their organisation's previous year's work and on its plans for the future.[5]

However, it was the growing attention to planning the future development of the Soviet Union which led to the initial steps towards a formal and comprehensive planning system. When, in 1926, the coordinating activities of KEPS were discussed at the First Gosplan Congress,[6] Krzhizhanovskii stated that Gosplan was not only interested in the correct organisation of the study of natural resources but also 'in the systematisation of the whole business of science'.[7] From the middle of the following year science was placed within the planning work of Gosplan. The first moves towards the planning of industrial research also came in 1927. In March a resolution of VSNKh's Presidium approved Sverdlov's report on the work of NTU.[8] It had three paragraphs dealing with the planning of the industrial research network. Firstly, it was recognised as necessary that *glavki* and trusts participated in the compiling of the institutes' annual research programmes, thus making official policy what was already happening in some fields. Secondly, NTU was to provide the Presidium of VSNKh in succeeding years with a plan summarising the intended work of its institutes; this was to be considered by VSNKh at the same time as the overall plan for industry. Thirdly, NTU was to undertake some detailed planning of the logistical support of research; for the VSNKh Presidium, while approving NTU's proposals for founding three new institutes, asked it to produce a detailed programme for their formation with material on administrative structure and plans of work, with financial estimates. In addition to increasing NTU's planning role this resolution also laid the basis for a possible further increase in the control over the direction of research. It proposed that sums of money should be included in the financial plans of trusts and enterprises which had to be spent through contracts (*dogovory*) with research establishments on specific research projects or in payment for consultation.[9] Grants from *glavki* should also in future be tied to particular pieces of work. In October 1927 the planning and coordination of the work of NTU's research establishments was one of the tasks given to each NTS.[10]

THE DEBATE ON THE PLANNING OF SCIENCE

This growing interest in the planning of science by central government organs such as Gosplan and VSNKh was accompanied by discussion of a more general nature on the application of planning to science, primarily initiated by Gosplan and marked by the publication of several articles in the Gosplan journal and elsewhere. Their authors generally agreed that in spite of its complex nature science could and should be planned. As one Gosplan worker wrote in 1930: 'Allotting such an important place to science and technology in carrying out the plan, we cannot leave research and technical studies outside the general system of planning.'[11] On the other hand there was some disagreement over planning methods – over the question 'who plans?'. The pro-planning scientists would appear to have seen the role of the central planning organs mainly as coordination, while planning workers foresaw a greater degree of central direction. This is brought out by two articles published in 1929; the first, in *Izvestiya* on 26 June, was written by Bakh; the second, by Vangengeim and Troyanovskii, two planning workers, appeared in *Nauchnyi Rabotnik*, the journal of the scientists' trade union, in November.[12] Bakh considered that the basis of planning was the compilation of plans by the research establishments themselves; planning cells in them would produce workplans which would be coordinated through annual conferences for the various branches of science. Bakh's main fear was bureaucratic direction from the centre. Conversely the Gosplan workers feared the consequences of giving too much independence to the institutes in planning matters. They agreed with Bakh on the need for primary cells to play a role in the planning process, but thought that the process should be initiated from the centre. They pointed out that science was not fully 'sovietised' and that planning was linked to ideological control over science. They further criticised Bakh for dealing solely with the planning of research topics; yet, they pointed out, finance, administrative control and manpower training were all centralised. These two different approaches were later clearly differentiated in a further article by Troyanovskii. He said that there were two ways to approach the planning of science, either through the 'statification' (*ogosudarstvlenie*) of science or its 'socialisation' (*obobshchestvlenie*) – in the latter case planning would be left to the scientists themselves. He still thought that some central direction was necessary and that the scientific plan could not fulfil the natural tasks of a plan if it was simply based on the coordination and review of lower-level plans. A global plan for science had to put forward its own problems and projects for wide discussion in scientific conferences and congresses.[13] However, there would appear to have been no general agreement even among planners on the optimal extent of central direction. Another writer in *Planovoe Khozyaistvo* seems to have been prepared to ascribe a somewhat greater role to the individual scientists than Troyanovskii; but mainly because he considered that it was necessary in certain circumstances to go so far as planning the methods used in research and that success in this field would depend on the widespread participation of research workers.[14] He also believed that some parallelism could be useful,[15] while for Troyanovskii it was 'never justified, neither from a scientific nor an economic viewpoint'.[16]

However, this detailed discussion of the form of planning certainly did not mean that the principle of research planning was quickly and universally accepted by scientists, as the authors of the articles themselves noted.[17] The articles were indeed part of an attempt to win support for the principle of planning. This was itself closely connected with the general campaign for linking science to economic development, which saw the publication of articles on the role of science in development[18] and also the organisation of ideologically sympathetic scientists into a body to spread propaganda among their colleagues. This was the All-Union Society of Workers in Science and Technology for Assisting Socialist Construction (VARNITSO), set up under the chairmanship of Bakh in 1927.[19]

The climax of the drive to introduce the idea of planning into science came in April 1931 with the holding of the First All-Union Conference for the Planning of Scientific Research. It was organised jointly by NIS VSNKh and Gosplan and had over 1000 participants. Its proceedings were widely publicised in the press.[20] The keynote speech was a three-hour address by Bukharin in which he presented a detailed account of the aims, possibilities and problems of the planning of science.[21] The report of the subsequent discussion in the newspaper *Ekonomicheskaya Zhizn'*[22] emphasised the need for scientists to completely change their attitude to plans; it complained of the lack of any concrete proposals or detailed discussion from the floor. The twenty or so speakers went no further than a declaration of the necessity for a plan for scientific work and some didn't even go this far.[23] Later in the conference Kuibyshev, in his speech as head of Gosplan, presented the participants with an extensive list of priority areas for future research.[24]

The holding of this conference was to signify the end of the campaign to introduce planning into science. At another conference on research planning just over eighteen months later, in December 1932, Bukharin stated that the idea of planning had been 'adequately put into the heads of research workers',[25] and subsequently much less emphasis was given to the general application of planning to science.

THE FIRST FIVE YEAR PLAN AND INDUSTRIAL RESEARCH

Within VSNKh's research network there was a much more ready acceptance of the principle of planning than in the Academy of Sciences or by scientists working in the establishments of Narkompros.[26] As we have seen, conferences were being regularly convened to coordinate research and the central review of institutes was being undertaken in the middle of the twenties. The applied nature of the vast amount of the research of the VSNKh institutes was more amenable to planned direction than the more fundamental research done by many research establishments elsewhere. In the more 'capital-intensive' fields of applied research there was also a greater need for institutes to bid for resources from the central organs.

VSNKh was playing an extremely important role in the work on the five year plan, and in November 1927, when it was feverishly working on its third draft for this plan in anticipation of the next month's Fifteenth Party Congress, there was a discussion within NTU involving leading scientists from its research network which resulted in a decision that VSNKh's research institutes should each draw up a plan for their work for the corresponding period.[27] At the start of 1928 it would appear that NTU itself began work on an overall plan for the development of industrial research.[28] It was a time when, in the aftermath of the great attention paid to the Plan at the Fifteenth Party Congress, work on it was greatly intensified. This growing activity supplied one of the stimuli for the review of the industrial research network in those months which led up to the comprehensive Sovnarkom decree of the following August.[29] Sovnarkom asked VSNKh to accelerate and improve the work on the five year plan for research. It also set out what it considered ought to be the aims of the future years' research programmes: these were the strengthening of the defence capability of the USSR; the release of Soviet industry from its dependence on foreign industry; problems concerned with the technical reconstruction of the country – electrification and chemicalisation were cited as examples; the development of new products and processes; and, in addition, work on general theoretical problems. When completed, this plan of research programmes was to be scrutinised by a congress of scientists and industrialists from VSNKh and other departments. Further, it was envisaged that when revised and approved this plan would provide the core of an integrated plan for the development of the industrial research network over the next five years. The distribution of resources – not only to the industrial research institutes, but also for work at the laboratories of educational establishments, in institutes of other departments and at factory laboratories – and the development of new facilities would be based on this plan. Thus the Soviet government was now to embark on the production of a bold comprehensive plan for the development of industrial research, with the available resources geared to the research programmes which were considered the most important.

In response to the Sovnarkom resolution VSNKh in its turn ordered its *glavki* and committees immediately to select from their five year plans the basic problems which needed investigation and which were linked to the general areas listed by Sovnarkom. This information was to be sent to NTU by the first day of the next month. The latter, in conjunction with the NTSy, was then to draw up a summary five year plan, the main problems of which were to be discussed at the proposed congress. On the basis of the suggestions of the *glavki* and the discussion at the congress NTU and the NTSy were to review the individual institutes' five year plans and draw up a plan for the development of research facilities on the lines proposed by Sovnarkom.[30]

The VSNKh decree did not set a deadline for NTU's presentation to the presidium of a draft of the five year plan. However, a newspaper report at a time of intensive planning activity within VSNKh suggests that deadlines had been set which NTU had failed to keep; for the issue of *Torgovo-Promyshlennaya Gazeta* for 24 November 1928 refers to a meeting of VSNKh's permanent planning commission at which Shein, deputy chairman of NTU, confessed that the drawing-up of the plan was somewhat delayed.[31] He said that it required further work and that a particular fault at that time was its failure to cover satisfactorily the necessary expansion of work in the provinces. Zolotarev, the member of the Presidium who chaired the meeting, gave NTU until 15 December to produce the control figures for the next five years. Yet Lapirov-Skoblo, in an article published at the beginning of 1929, said that it was proposed to finish the five year plan for the development of research facilities in February.[32]

It would seem that the estimates of research expenditure which were being worked on at this time provided the data which appeared in mid-1929 in an article by P. S. Osadchii of Gosplan. This article brought together a whole series of departmental estimates for research expenditure over the plan period.[33] Total expenditure on research institutes and establishments under NTU would be 359 million rubles

- 154 for capital expenditure and 205 million current expenditure.[34] It was foreseen that the institutes would get about one quarter of their funds (95 million rubles) from non-budgetary sources.

Sovnarkom and VSNKh had both envisaged that the plan for the development of research facilities would be based on the plan for research work. The planning process was also not proceeding smoothly in this area. It too was unfinished at the beginning of 1929. Lapirov-Skoblo expected it to be ready by March and therefore after the plan which was supposedly based on it. According to Lapirov-Skoblo the research plan was not simply going to list topics but was to state the time the work would take, its approximate cost and the source of funds for it. NTU was said to have approached several other commissariats and VSNKh's Glavvtuz with requests for their research plans for the purpose of coordination.[35]

A further part of the work on the plan was the compilation of five year plans by NTU's research institutes. At least twenty-five institutes produced plans which were later published as pamphlets.[36] These were mostly between twenty and fifty pages in length and contained projections for manpower and expenditure and a list of intended research projects.[37] It was these plans which were presumably to be amended on the basis of the overall plan which NTU had been ordered to draw up for the research network as a whole, for NTU's expenditure estimates would appear to have been largely independent of those produced by the institutes. The addition of data from the twenty-four available plans gives an expenditure of 392 million rubles compared with the 359 million total cited by Osadchii. According to his data a quarter of the funds were to come from non-budgetary sources, while the proportion was 40 per cent in the institutes' plans.[38] On the other hand, NTU would appear to have drawn up no separate data for the future expansion of research manpower; it was left to the institutes to determine their own requirements on the basis of their proposed programmes for research. For in an article on scientific manpower published at the end of 1929 Lapirov-Skoblo reported that no comprehensive data existed on future manpower needs, and the figures he cited were stated to be the projections of NTU's institutes. On this basis the number of scientists was to grow from 2000 to 3650 'in the course of the current year' and to 5574 at the end of the five-year period.[39]

Any consistency that may have been achieved in the work on the plan between NTU and its institutes in the period from August 1928 to April 1929 was undoubtedly to disappear in the succeeding months.

When Sovnarkom approved the adoption of the optimal variant of the First Five Year Plan on 23 April, it also gave a directive for certain changes to be made. One of them was that the expenditure on industrial research should be increased by not less than 150 million rubles.[40] Furthermore, the industrial research network was expanding at a rapid rate as a result of the stimulus given to research by the Sovnarkom resolution of the previous August. The number of scientists had reached almost 5000 by the beginning of 1930.

The lack of any detailed work on a plan for research manpower was becomingly increasingly apparent, for there was a growing and serious shortage of qualified manpower throughout the economy. This problem was discussed at a plenum of the Central Committee in November 1929. The resulting resolution spoke of the lack of soundly based plans for manpower training and noted the short supply of various specialists needed in industry and research. Gosplan and the commissariats were to produce a new five year plan for the training of specialists.[41] In fact the redrafting of the plan for the training of specialists became part of a general revision of the five year plan which began at the start of 1930.[42] In July, as a consequence of Gosplan's directives, VSNKh initiated a reworking of the plan for the industrial research network.[43] While in 1928 it had been envisaged that NTU would itself construct the plan for the development of the research network on the basis of the programme for research approved by the proposed congress of representatives of science and industry, the emphasis was now on the *ob"edineniya* which controlled the separate branches of industry. On the basis of directives sent out by VSNKh each *ob"edinenie* was to start an immediate review of the plans of its subordinate institutes, ensure that plans were drawn up for the work of laboratories at trusts and enterprises and produce a plan reflecting the future development of research and research facilities in their branch. NIS was to produce a summary plan on the basis of this series of plans. It was intended that this whole process should be over by the end of October 1930. However, NIS had certainly not acquired all the necessary plans by the time of the planning conference in April 1931, and may never have done so.[44]

Thus, the attempt in the Soviet Union to produce a comprehensive plan for the development of applied research linked to industrial development foundered in the hectic activity of these years. It suffered from the same process of change and reappraisal as the plans for other parts of the economy. The seemingly general failure of any estimates to be produced by the established deadlines also reflected the low priority attached by industrial organisations to matters of science as opposed to matters of production and probably the initial anti-planning attitude of many scientists. An additional factor after 1929 was the decentralisation of the research network and NIS's lack of power and authority in the early years of the 1930s. This latter development was also instrumental in the completely changed nature of the revised plan that was to be produced in 1930. The devolution of planning responsibility to the branch level which took place at this time was to be maintained. No further attempt was to be made to draw up centrally a comprehensive plan for directing the future development of industrial research on the lines embodied in the Sovnarkom resolution at the beginning of work on the five year plan. Overall plans were in future to be based on the addition principle with the straightforward incorporation and summarising of branch plans.

The work on the plan for industrial research was the major planning activity concerning science during these years. In this respect it was a reflection of the activity of VSNKh as a planning organ; the five year plan as a whole was as much a creation of VSNKh as of Gosplan. Gosplan played little part in directing plans for the development of science. The figures which it produced at the end of the 1920s were based on the addition of estimates forwarded to it by departments. The published plan contained only scattered information on the development of science and data on the growth of certain parts of the research network, which was clearly the data forwarded to Gosplan by the department concerned.[45] In fact throughout the 1930s Gosplan appears to have continued to play the role of a summariser of plans submitted from below rather than that of a director of research programmes from above.[46]

The branch basis on which the planning of industrial research was implemented was illustrated by the procedure for compiling a list of research topics to be undertaken during the Second Five Year Plan. The plans compiled by individual institutes during the course of 1932 were scrutinised at the Second All-Union Conference for the Planning of Research in Heavy Industry, which was convened in December 1932 by NIS, the Academy of Sciences and Gosplan to undertake a detailed review of the proposals for research in the second plan and in 1933. This conference of representatives of science and industry, working on the basis of sections dealing with the various fields of research, produced a detailed list of intended research projects for the next five years and some suggestions for the development of research facilities.[47]

THE ANNUAL PLANNING OF INDUSTRIAL RESEARCH

The core of the research planning system, as in the case of the economic planning system as a whole, was the annual plan for each research institute. The annual review of workplans which was being undertaken in the mid-1920s was the initial step towards the development of a comprehensive annual planning system. The VSNKh resolution of March 1927 extended these activities and also envisaged the linking of some funds to projects through contracts on the customer-contractor principle. Similarly, the Sovarkom resolution of August 1928, which was to be responsible for the start of work on a comprehensive five year plan, was an important stage in the development of annual planning. It instructed VSNKh to provide Gosplan with a summary plan of the research to be undertaken by NTU's research establishments in 1928/29 and similar annual plans in the following years.[48] The subsequent VSNKh decree ordered NTU to present *glavki* with a plan of the work of its institutes by 25 September, i.e. within five days. A revised draft was to be produced on the basis of criticisms made by industry. The plan was also to be coordinated with establishments in other departments. At the beginning of 1929 Lapirov-Skoblo wrote that it was proposed to finish work on this annual plan for research in March.[49]

Work on the plan for the next year does not appear to have got under way until after the start of the 1929/30 year. A decree of VSNKh published in mid-November[50] critically reviewed the previous year's initial attempt to produce a research plan, with the aim of pointing out the changes needed in the current year's plan. The main fault, it was stated, had been its limitation to NTU's research institutes and a lack of information on the work of trust and factory laboratories. There had also been nothing on the specific purpose of research projects nor any measures directed towards achieving a better regional distribution of research.

In the following year the scope of the annual plans was widened to include research funding, though only current expenditure.[51] Several factors were behind this development. It was obviously a natural step towards fuller central planning of research. But it was also a corollary of the overall linking of science to the industrialisation programme, which was resulting in an expansion of non-budgetary finance in general and in particular of funds tied to specific projects and, perhaps, a necessary measure to counteract the consequences of the changes in the industrial research network itself; since research institutes were

Research Planning 89

increasing in number and were now controlled by a number of different organisations (as of August 1931, NIS and thirty-nine *ob"edineniya* and trusts), there was a greater need for a systematic plan than when there were a few institutes all grouped under NTU. Such a plan would also help to ensure that the industrial organisations paid the necessary attention to the research institutes and also to the laboratories of trusts and factories; for the role of the latter as research organisations was now being increasingly emphasised. The guidelines for drawing up the plan for 1930/31 put the expansion of such laboratories at the top of the list of priorities.[52]

These guidelines accompanied a VSNKh decree of 11 May 1930 which initiated work on the next year's plan.[53] In many respects they foreshadowed the guidelines for the reworking of the five year plan which was to begin two months later. VSNKh gave the organisations which controlled the research establishments until 10 July to submit the plans to NIS. However, this deadline, like so many others, was not met and eleven days after it expired VSNKh issued a sharp decree which stated that the situation was 'absolutely impermissible' and that the failure to produce the plan was 'disrupting the planning work of VSNKh and Gosplan.'[54]

While the 1930/31 plan was to be the first to incorporate financial information, it was as we have seen limited to the institutes' current expenditure. NIS was not given the authority to plan capital expenditure until February 1931 – a measure proposed in the resolution of VSNKh, which aimed at increasing its general authority over industrial research. The plan for 1932 was, thus, the first to cover fully all aspects of research. The information which the research establishments were to provide was by now extremely detailed. Each project had to be accompanied by such information as the duration of work, the number and qualification of the personnel involved, the cost, and even where the results would be used. All this material had to be supplied both to NIS and to the relevant *ob"edinenie*. On the basis of the individual plans each *ob"edinenie* was to present NIS with a summary plan for the institutes and laboratories in its branch of industry.[55] VSNKh appears to have determined at the outset of the work on compiling the 1932 plan to try to avoid a repetition of the previous years' delays. Five days after the publication of the plan guidelines it issued a decree on the role of economic organisations in the work on the plan, which stated that the research institutes were to report to the Presidium of VSNKh any failure by them to fulfil their obligations.[56] The 1932 plan was, in the event, completed before the start of that year. Ziskind of NIS reported in December that he had reviewed two hundred plans in the previous month[57] and an article by him summarising the plan had appeared by the end of 1931.[58] Similarly an article on the 1933 plan appeared in the last issue of NIS's journal for 1932,[59] but this plan must have undergone considerable revision following the reorganisation of the research network which was just about to be implemented.

Thus by the end of the First Five Year Plan an annual planning system had evolved under which institutes were compiling plans giving detailed information on the programmes they intended to undertake and the resources which they would use on the work. The planning process was started by the issue by the central apparatus of the commissariat – NIS until its abolition – of a document which listed the priority areas for research in the coming year.[60] Within the institutes a plan would be prepared by the planning departments which had been established in each institute by the mid-1930s in collaboration with the institutes' research scientists. When these had been approved by the various institutes' administrations, they were reviewed in conjunction with plans for research in the same field at meetings in which representatives of industry took part. In the mid-1930s this review was undertaken by research associations for the various fields. A few such associations had existed from the very end of the 1920s but the majority were set up in the aftermath of the First All-Union Conference for the Planning of Scientific Research; they played a role in research planning similar to that entrusted to VSNKh's NTSy in the late 1920s.[61] These associations were abolished in 1936 and this review of research plans was subsequently carried out by *ad hoc* committees.[62] At these branch meetings considerable pruning of the plans could take place.[63] The plans revised in the light of these changes were submitted for approval to the central apparatus of the commissariat. When this had been obtained, the institute worked out detailed control figures (information on costs, manpower needs etc.) for each project. When these were passed, the institute produced a still more detailed programme for each project which included quarterly and monthly subplans. J. G. Crowther reports that in Ioffe's Leningrad Physical Technical Institute the preparation of plans began at the end of October and that they were ready at the end of December.[64] Elsewhere, however, we find criticism of delays in approving the research plans of institutes and in the production of a summary plan by the central authorities.[65]

THE FINANCE OF RESEARCH AND THE CUSTOMER-CONTRACTOR PRINCIPLE

An important step towards the development of a detailed annual planning system and, perhaps, one of the reasons for its development was a radical change in the early 1930s in the method by which research was to be financed. From 1927 funds could be tied to specific projects by industrial organisations making a contract for a piece of work with an institute, and in 1930 and 1931 around 15 per cent of the funds which VSNKh research establishments enjoyed were on this basis.[66] In 1932 there was a massive switch to this method of funding.[67] This was the result of the application of *khozraschet* (economic accounting) to industrial research as part of an emphasis on the full introduction of *khozraschet* throughout the economy.[68] The term *khozyaistvennyi raschet* or *khozraschet* had originally signified independence from the state budget and meant that economic units had to finance their own production out of sales with the aim of earning a profit.[69] Later it became closely associated with cost-cutting and economic efficiency.[70] When applied to the research institutes it was to mean that they would obtain their funds not through direct subventions from the state budget but by selling their 'production' – research results – to industry. Applied research was in future to be done on the basis of extremely detailed contracts (*dogovor*), or orders (*naryad-zakaz*) where the agreement was between an institute and the *glavk* or *ob"edinenie* which controlled it. It was considered that this mode of finance would strengthen the research establishments' links with industry, raise the quality of research and improve its cost-effectiveness. Institutes were to be paid as they completed the various stages of the work and receive a bonus if the expenditure limits were not exceeded. Contracts for research on problems connected with improvements in existing production technology would be financed from current production expenditure, work on new processes and products from capital expenditure allocations which were earmarked for research and which had previously been handed over to the institutes in the form of block grants.[71] This system of finance proved to be very detailed and unwieldy, especially when the principles of *khozraschet* was extended to the individual departments of research establishments.[72] There was to be growing criticism of the finance of research through the blanket use of contracts, firstly, on the grounds that it resulted in great pressure on the institutes to provide the immediate scientific servicing which the plants should themselves provide from their own scientific facilities;

and secondly, that the lengthy and complicated procedures associated with the contracts and the attitude of industry resulted in institutes having no reliable regular source of income.[73] There was a growing call for a degree of more centralised finance which resulted in the decree of October 1936 completely revising the method of finance for research in heavy industry.[74] It was no longer to be based on the contract principle. Finance was now to be concentrated in each *glavk*. While previously NIS NKTP had allocated the research funds from the state budget to be used for theoretical or inter-branch work directly to each institute, it was now simply to be responsible for distributing the funds among the *glavki*. Capital construction funds through which funds for research had formerly been allotted were now only to be used to fund research connected with the design of new plants. The main source of research funds was in future to be the allocation to a central fund for research within each *glavk* of a proportion of the sales revenue of the particular branch; the actual size of this proportion is unknown, but it does appear to have been the same percentage in each branch of industry.[75] Only work on the improvement of existing production was, as before, to be financed through contracts between institutes and enterprises; the funds were to come from the plants' current production expenditure.

The situation in which the *glavk* was to be in charge of the disposal of research funds was, in fact, subsequently to be criticised on the basis that the system of deriving funds from a proportion of sales revenue meant that in some branches of industry, where output was small, research funds were limited, while other *glavki* were awash with funds for which they had no good use.[76] However, it remained the basis for research funding until the outbreak of the war. This return to a more centralised form of finance did not mean a return to the previous annual block grants – funds continued to be tied to particular projects and payments continued to be made in stages – but it did deal with the two major criticisms of the contract system by ensuring a more reliable supply of funds to the institutes and also by lessening the pressure on them to act as the enterprises' scientific laboratories.

SUCCESS CRITERIA

The development of suitable criteria for measuring success and efficiency in research was clearly necessary for the successful application of detailed annual planning to industrial research, particularly in any attempt to incorporate some degree of payment by results for R&D personnel. The major problem in assessing performance in scientific research was (and continues to be) the measurement of scientific output. In the Soviet Union during the development of the planning system there was some suggestion that meaningful figures in money terms could be obtained for the assessment of some aspects of applied industrial research. S. G. Strumilin, the economist and planner, considered that the effectiveness of R&D on new production processes could be satisfactorily measured by the resulting savings in production costs.[77] The guidelines for drawing up the 1930/31 industrial research plan foresaw a role for some measurement of effectiveness both in the selection of projects for inclusion in the plan and in assessing completed work.[78] At a conference on planning within research institutes which was held in December 1931 Ziskind of NIS criticised the plans produced by the industrial research institutes for lacking any clear information on the expected results of research. He considered that estimates of potential effectiveness were necessary to 'direct the work of the researcher'.[79] However, there was no general approval for the use of economic effectiveness as a criterion in assessing research performance. While it might have been possible to use it as a measure of the success of newly-developed processes, there was no likelihood – nor, in fact, suggestion – that a way could be found to use such an approach for all industrial research.

In fact, Sul'kevich, a colleague of Ziskind at NIS, also attacked the use of economic effectiveness as a measure of the quality of research on the grounds that quality was determined not only by results and that negative results did not always mean that the work had been of low standard.[80] In making this distinction he reflected both the view to be expressed by Strumilin in a booklet which was to appear in the following year and that of an earlier writer, N. P. Suvorov.[81] Both these authors discussed in detail the possibility of developing a comprehensive synthetic criterion for measuring production and success in the various fields of scientific research, by weighting the output of research publications to take account of their 'internal' scientific quality and their 'external' quality – in simplest terms their contribution to the development of the Soviet Union. Both Strumilin's and Suvorov's work was basically an academic exercise. However, similar approaches were being suggested for actual use in industrial research. At the December 1931 conference Sul'kevich referred to proposals for criteria for measuring research and assessing plan fulfilment which involved the calculation of a synthetic unit of research output, which not surprisingly he considered to be impracticable.[82]

In complete contrast to the use of complicated weighting procedures were suggestions that scientists should simply be judged on the basis of the number of pages produced in a year. It is reported that one (unnamed) institute proposed to regulate the pay of its personnel on the basis of their output of literature.[83] This establishment suggested that only half a salary should be paid automatically, the rest would depend on output. Such an outright quantitative measure was unacceptable to the authors of the article, as it had been earlier to Bukharin, who at the First All-Union Conference on the Planning of Science spoke of the inadequacy of crude measures based on number of publications or quantity of printed pages.[84]

Nevertheless, the criteria which were actually used by the research network took no account of the varying quality of work. Methods in use in the early 1930s were elaborations based on comparisons of the actual number of man-hours or man-months needed with those envisaged in the plan or on the proportion of planned projects which had been completed.[85] The difficulty of producing meaningful output figures for research resulted in the use of such input data as man-hours becoming the general basis for assessment of plan fulfilment. Efficiency was measured not by a comparison of expenditure with output but by comparing actual and planned expenditure. Thus, with the introduction of *khozraschet*, a source of bonus funds was to be the savings achieved on planned expenditure for contract work, and scientists were also rewarded for spending less than planned on projects financed directly from the state budget.[86] However, the main source of bonus funds for research staffs would appear to have been the allocation to the bonus fund of a proportion of the institute's wage-bill (1 per cent in the early 1930s).[87] Although no information has been found on the proportion of a research worker's pay which consisted of bonuses for fulfilling his research plan, it was probably only a marginal part of his remuneration.[88]

THE SUCCESS OF PLANNING IN SOVIET INDUSTRIAL RESEARCH

Thus over the years of the late 1920s and early 1930s a system of research planning was developed in the Soviet Union and by the outbreak of the Second World War research planning was a well-established fact of life. How successful was this system in operation?

The comprehensive planning of scientific research basically has three aims: firstly to ensure that the research projects which are undertaken are the right projects; secondly to provide the logistic support of research and to match research projects with resources; and thirdly to ensure that work on these projects is carried out in the most timely and efficient way. The available evidence enables us to make a tentative evaluation of the success of planning in these three fields.

Project Selection

To measure fully the success of planning in project selection would require a detailed branch-by-branch analysis of the state of Soviet industry and of the developments envisaged in the economic plans, to be followed by an assessment of the extent to which the planning system concentrated work in those areas and on those problems of greatest importance for carrying out this industrialisation programme. With regard to the more theoretical work, assessment would be based on the degree to which such research was concentrated in those fields which, with the benefit of hindsight, proved to be the most fruitful in producing results which were to lead in time to applied research projects and developments of industrial importance. Such a detailed approach is beyond the scope of this book; and our material is limited to the question of the success with which the planning system ensured that research met the needs of the economy as seen by party and government.

The introduction of planning was undoubtedly to help in the process of directing the research effort towards servicing the industrialisation programme. This reorientation of research was seen to depend not only on research plans being drawn up in line with general central government priorities, but also on the establishment of an efficient mechanism for industrial organisations to initiate more specific research problems. There was, however, to be continuing criticism of the extent to which institutes' research projects neglected the more specific needs of industry, and both institutes and industry itself were blamed. The *ob"edineniya* and later the *glavki* were to be often taken to task for failing to exercise the necessary control of the plans of the institutes or for not providing the institutes with the necessary information.[89] Institutes on the other hand were attacked for failing to reflect in their plans the problems which had been suggested to them by industry.[90]

The introduction of the customer-contractor principle and *khozraschet* into the annual planning of industrial research was an important factor in the drive to improve these links between research institutes and industry. It resulted in a rapid growth in the proportion of research funds tied to particular projects. However, the actual effect on project selection can, perhaps, be overstated. Industrial organisations were not too willing to fund research and were likely to want to use the funds allotted for contract research on low-powered projects of immediate application in production. Bukharin in 1932 had spoken of a danger that institutes might be turned into 'bad rationalisation offices'.[91] Further a consequence of regulations which forbade institutes to sell results produced on a contract for one body to other organisations was that some industrial bodies were loth to finance work, hoping that a related organisation would provide the funding and that they would then acquire the information free of charge.[92] On the institute's side the growing proportion of its total funding which was coming from contract research may have led in some cases to contracting for work which was peripheral to the institute's main interest, simply of financial necessity.

In spite of the establishment of planning departments in institutes, the extent to which they produced plans of research projects that were in line with central government demands and the suggestions of industry continued to depend largely on the attitude of the research scientists themselves. Planning staffs particularly in the early years are reported to have concerned themselves mostly with the more quantifiable areas of finance and manpower.[93] Consequently there appears to have been no pronounced shift towards the production of a plan based on a coherent and interlinked set of projects. Scientists probably continued to work to a large degree in areas which interested them. A frequently noted result was the phenomenon of *'mnogotemnost''* – the inclusion in the plan of more projects than the institute's resources could properly cope with.[94] Scientists may also have dressed up titles of projects in a form more likely to be approved as fitting in with the requirement to emphasise work of use in the industrialisation programme.[95]

Clearly an important function in producing a coherent plan of research rested with the research associations and the *ad hoc* meetings which replaced them. It is impossible to assess their effect on the plan, but an important handicap would have been their lack of authority, for they did not have the power to reject plans outright or order changes. In fact at the beginning of 1934 Academician Bakh had suggested that the research associations should be given more teeth by the establishment of a strong government organ which would operate on the basis of the activities of the associations and have the full authority to coordinate and plan research.[96] A reflection of the position of planning in the first half of the 1930s is that NIS would appear to have borne a large proportion of the actual responsibility for coordinating research programmes in spite of all the other mechanisms which were in existence. One of its staff, E. Romanovskii, wrote in 1934 that in his work at NIS over the years since 1930 he had frequently come across huge distortions in the formation of research programmes which stemmed from the lack of knowledge on the part of scientific establishments and planning organs of research being done outside their department, city or institute.[97]

The planning of the logistic support of research
The projections for research funding and manpower which were the result of the work on the first five year plan for research were to show a wide variation from what actually happened. In mid-1929 Osadchii had given 359 million rubles for the total expenditure on research under VSNKh; in the four and one quarter years from October 1928 to the end of 1932 somewhere around 600 million rubles were spent.[98] The manpower estimates quoted by Lapirov-Skoblo at the end of 1929 also severely underestimated future growth – the figure for scientists which he cited, 5574, had been exceeded by the start of 1931.[99] Such comparisons, however, should not be made in isolation, for the revisions and changes in the five-year plan for research and the actual growth of the network reflected the changes in tempo and policy which affected the five-year plan for the whole of the economy.

The first comprehensive annual plans for the expansion of research were also greatly to differ from actual developments. The clearest example is provided by the NIS NKTP control figures for the commissariat's research establishments in 1933 which were published at the end of 1932. These did foresee some pruning of the research manpower, from 45,000 at the beginning of 1933 to 42,000 at the end, but this substantially underestimated the severity of the impending reorganisation. The total employment on 1 January 1934 was under 35,000. A planned rise of 400 in the number of scientists became a drop of 4000.[100]

The extent of the central coordination between the plans of the resources for industrial research and projected research programmes was probably small. The original five year plan for manpower and funds was to have been based on the plan for research projects, but, as we have seen, in practice this proved to be an impossible task. Later, in

1934, Kviring, a deputy chairman of Gosplan, wrote unfavourably of the planning of science in that organisation; particularly that its sector for science was chiefly involved only with the funding of science.[101]

Thus, it was largely in the research establishments that the financial, manpower and project plans were brought together. The introduction of the customer-contractor principle tied a very large part of their funds to specific projects. Further, the planning system involved the construction of detailed plans which linked research programmes to manpower and financial resources. However, these plans were basically used to allocate the funds acquired by the institutes among laboratories and projects. The workplans which were put forward at the start of the planning process were not accompanied by data on the cost of labour needs. In view of the remarks that the research establishments were taking on more projects than they could cope with and that their planning personnel were also in the main concerned with the allocation of resources, it is likely that in fact no general systematic coordination of project and resource plans was being undertaken within the institutes.

Planning and efficiency

The third aim of research planning is to ensure that work on the projects in the plan is carried out in the most timely and efficient way. In the period under review the failure to develop a usable set of meaningful criteria for judging performance prevented planning from exerting a substantial influence on the cost-effectiveness of research in Soviet industry. The success criterion was simply the proportion of the workplan which was completed and financial incentives to research personnel were probably small. Furthermore, a failure to actually complete the research projects listed in the plan appears to have been a relatively frequent phenomenon throughout the period, and not restricted to the poorly-staffed and equipped institutes.[102]

In the early 1930s the introduction of the customer-contractor principle and *khozraschet* was seen as providing an incentive to improve the actual process through the phasing of payment over the period of work on a topic, by the withholding of a proportion of the funds until the satisfactory completion of the work and by the provision of bonus payments tied to savings on planned expenditure. These were all measures clearly designed to increase cost-effectiveness. However, success in this aim was hindered by the abuses of the contract system which led to the changes in the financing of research in 1936. Both industry and the institutes were attempting to bend the system to their financial advantage, thereby nullifying its proposed beneficial effects on research performance. For example, some industrial organisations apparently refused to make all the agreed payments; thus the chemical *ob"edinenie* signed a contract with the North Caucasus Institute for Applied Chemistry, but when the institute completed the work in half the specified time the *ob"edinenie* only paid 1000 out of the agreed sum of 3000 rubles on the basis that the work was not as complicated as predicted.[103] On the other hand institutes were contracting for work which differed radically from their plans, apparently concluding contracts on the same topic with several organisations (charging the full cost to each) and selling previous contract research results which should have been freely available to all industrial organisations.[104] In the second half of the 1930s institutes were also abusing the arrangement under which payments were made for the completion of each stage of work on the unverified statement of the institute itself; by declaring work to be completed to a greater degree than it actually was, the institute was acquiring extra funds.[105]

In another response to the planning system research establishments would seem to have acted in a similar manner to industrial organisations with an acceleration in the pace of work as the end of the yearly plan period approached. J. G. Crowther reports that such 'storming' took place in December in the Ukrainian Physical Technical Institute.[106]

A corollary of the introduction of a planned economy was, therefore, the application of planning to science. An important difference between the way industrial production was planned and the system for planning industrial research programmes was the role given to the 'grass roots' in research planning both in the institutes and through bodies such as the research associations of the first half of the 1930s. This can be contrasted with the overwhelming central direction of the rest of industry. This reflected the view taken by Bakh in the general discussion of approaches to planning science at the end of the 1920s. It was a recognition of the special position of research. As I. K. Luppol of the scientists' trade union stated at the First All-Union Conference for the Planning of Scientific Research: 'in the business of the planning of the topics of scientific research, the bureaucratic approach is completely unsuitable, for it is impossible to have in the bureaucracy a significant number of scientists who are doing research.'[107] It did however mean, as Vangengeim and Troyanovskii had hinted in 1929, that the systematic linking of research projects to resources was made more difficult, for the level of decentralisation was not matched by any devolution in planning the logistic support of research. At the same time, notwithstanding the initial bold attempt to produce a comprehensive five-year plan for the development of industrial research, the central organs of the industrial commissariats and Gosplan paid more attention to future developments in manpower and particularly funding than to producing integrated research programmes. On the other hand the emphasis on the customer-contractor principle in financing research did lead to the linking of funds to individual projects at the lower and more detailed level of the institute.

While the planning system did play a part in the direction of research into areas to which the government attached priority, the change in the pattern of industrial research was mainly the result not of developments foreseen in plans but of particular acts of government policy; the reviews and reorganisations of the end of 1932 and of 1936 are the most obvious examples.

8 Science at the Factory

The boost given by the First World War to the organisation and development of research led in the West to increasing discussion of the role of research laboratories and strong emphasis was placed on the need for organised factory research. Writers of the early 1920s[1] suggested that enterprises needed three types of laboratories: those carrying out analytical control over materials, processes and products; secondly, laboratories working on improvements in products and processes, and on new products; and finally laboratories investigating the fundamental sciences associated with the industry – research with no specific commercial object. With the development of a network of independent research institutes serving Soviet industry, the Soviet factory did not need to play such a comprehensive role in R&D and the scientific servicing of production. Clearly it was likely that fundamental research of interest to industry would be concentrated in such institutes, which would have an overview of the general problems of the development of a branch of industry. The main question concerning the role of factory facilities in the R&D system that was to be established in the Soviet Union was to be the extent of its involvement in applied research and the development of new products and processes. However, prime importance in the early years of the Soviet period had to be given to the establishment and restoration of laboratories to provide the necessary scientific control over production.

Factory laboratories were largely ignored during the economic exigencies of the immediate post-revolutionary years. Many were probably closed down completely and in the middle of the twenties Flakserman of NTO VSNKh singled out the lack of factory laboratories as the primary cause of previous poor links between research institutes and industry.[2] The restoration of industry with the consequent consideration of future expansion was as important a factor in initiating discussion on the role of science in the factory as in encouraging the general expansion of the industrial research network. In the years before 1930 this discussion was to be mainly concerned with the direct scientific servicing of production.

As a first step towards strengthening the role of science in industrial plants, in 1925 NTO, presumably at the request of VSNKh itself, undertook a survey of existing facilities, in the first instance in the plants of Moscow and Leningrad.[3] In the survey of Leningrad 140 factories were studied between September 1925 and May 1926.[4] The position of factory science appeared rather gloomy. Over half the laboratories investigated were found to be lacking some important requirement for normal efficient operation – premises, equipment, staff or funds. Only in a small number of cases was there any possibility of a laboratory undertaking scientific research. However, it was felt that the position of a substantial number of laboratories could be radically improved by factory managements simply taking an increased interest in them, at the basic level by providing a permanent and regular supply of funds. Yet the growing campaign for economy and the cutting of unnecessary expenditure in industry probably limited the extent to which resources for laboratories were spontaneously increased as a result of the attention paid to them by the commissariat; for it was reported that in the metallurgical industry 'the slogan of the "regime of economy" ... was misunderstood by many and this had the general consequence of cuts in the already scanty funds for laboratories'.[5]

Besides suggesting ways of improving existing laboratories, the commission which had undertaken the survey of Leningrad also recommended that twenty-five new laboratories should be established. A total expenditure of 1,800,000 rubles (equivalent to 14 per cent of expenditure on VSNKh's research establishments in 1925/26) was suggested to bring the overall level of the laboratory provision up to scratch.

NTO approved the proposals made by this commission in July 1926. It also decided to use the experience gained in Leningrad as the basis for extending the survey to cover the whole country. This larger study was started at the beginning of the following year. Discussing it Sverdlov made a detailed statement on the role of factory laboratories which was to be reflected in the policy taken towards them over the next years. He wrote:

> Efficient factory laboratories provide the necessary control over materials and the processes of production, provide for product quality, for the timely utilisation by industry of the achievements of the institutes, and in the reverse direction the correct formulation

Science at the Factory   103

and timely presentation to the institutes of those tasks the solution of which is within the capabilities only of central powerful scientific organisations, endowed with the corresponding apparatus and cadres of highly qualified scientists.[6]

Thus, the basis of VSNKh's policy was that rather than being strong research organisations, the laboratories in factories should act as the mediator between production and the independent research institutes, assisting the latter in introducing new developments and, on the basis of their knowledge of the production process, suggesting to the institutes the most suitable areas for research.

The importance of factory laboratories was now being very widely stressed. In 1927 reference was made to the need to strengthen the factory laboratories in the Central Committee resolution on the rationalisation of production,[7] in the resolutions of the Fourth Congress of Soviets on the development of industry[8] and of the Fifteenth Party Congress in connection with the five year plan.[9]

By 1928 there would appear to have been some improvement in the position of plant laboratories and in their vital work on production control. At the First All-Union Conference of Representatives of the Factory Laboratories of the Metal Industry, which was held in that year, reference was made to a new study of Leningrad which had produced evidence of the start of an important step forward;[10] similarly, in the Ukrainian metal industry a great growth in current production control work had occurred.[11] Although the main emphasis had been on control work and on the provision of the necessary equipment and qualified manpower, some research was being done both in Leningrad and the Ukraine.[12] In spite of this apparent improvement the NKRKI report in 1928 on the organisation of research for the needs of industry claimed that factory laboratories had still not received the development which met the demands of industry for science.[13] The decree of Sovnarkom which was based on this report and which, as we have seen, was an important landmark in the development of industrial research included several measures, specifically concerning the laboratories, which were to be implemented by VSNKh.[14] The existing laboratories were to be improved and had to become the main bodies for taking up the scientific and technical ideas of industrial workers. The five year plan for developing trust and factory laboratories was to be coordinated by the *glavki* with the future development of the institutes under NTU. Official regulations setting out the organisational structure and activities of factory laboratories

104 Science and Industrialisation in the USSR

were to be produced as quickly as possible. They should include a strengthening of the role of the laboratories in the production process. Lastly, in future financial plans the trusts were to provide for the necessary funding of the work of the laboratories.

The regulations for organising laboratories to which Sovnarkom referred were approved by the Presidium of VSNKh in May 1929.[15] The factory laboratories were seen as having four tasks. They were to do control and testing work on raw materials and products, undertake research which would prepare the path for the future development and improvement of production, establish links with the central laboratory of their trust, with research institutes and the laboratories of higher educational establishments, and control the factory library. Among a list of twenty-two rights and duties (*prava i obyazannosti*) were mentioned: the study of methods for the quality control of semi-fabricates and finished products, the testing of new processes both at laboratory and pilot-plant scale, and joint work with other scientific and technical establishments on questions linked with improvement in technological processes and the organisation of experimental work aimed at introducing the results of such activity into production. While individual sections of the regulations did foresee that rather than referring all possible research projects to the institutes, as suggested by Sverdlov two years earlier, laboratories would themselves undertake some research work, the main emphasis remained on the development and expansion of control and testing work.

The increasing attention to factory scientific facilities appears to have resulted in a rapid expansion in the numbers of factory laboratories in the years 1928 to 1930.[16] At the same time it is reported that financial support remained poor.[17] Part of the blame was laid at the door of the central organs. NTU was criticised for its failure to cover fully trust and factory laboratories in its initial work on planning the research network.[18]

Continuing dissatisfaction was probably the prime reason for a renewed survey of plant laboratory facilities in 1930. A sample of twenty-two large factories was selected and VSNKh's research institutes were to undertake comprehensive studies of factories in their particular field. The aim of the survey was to find out the role the laboratories played in the work of the factories, their funds and manpower, the state of planning and their long-term potential. The institutes were to produce specific proposals for future improvements.[19] This new study, like its predecessors, revealed widespread faults in the organisation and operation of the laboratories at enterprises.[20] Personnel had to be improved; urgent action was needed to provide equipment and facilities; half the laboratories studied had no plan. VSNKh proposed measures to deal with these faults and stressed that all laboratories should be organised in line with the published regulations and that all laboratories within an enterprise should be amalgamated into one unit under the factory director. Significantly the laboratories were also 'to strengthen their research work'.

Indeed from 1930 factory laboratories were increasingly to be seen as bodies which should do research. A VSNKh decree of June 1930 on the procedure for organising research establishments had included trust and factory laboratories under this heading.[21] Factory laboratories were to be looked on to a growing degree as the lowest part of a unified system of research establishments, of which the other parts were the research institutes and the laboratories of educational establishments.[22]

Decisive improvement in the laboratories as a result of the 1930 survey and the subsequent VSNKh decree was probably greatly hindered by the continuing process of industrial reorganisation and a consequent failure to establish a firm structure for factory science. The deadline set by VSNKh for reports on the implementation of the measures outlined in its decree of 7 September was not met.[23] A resolution of the Presidium of VSNKh in February 1931 instructed NIS to take 'all necessary measures' to improve and expand factory laboratory facilities and the *ob"edineniya* to make full administrative provision for such bodies.[24] Nevertheless the number of factory laboratories and the scope of their work was growing. According to a 1931 survey of scientific establishments which specifically excluded laboratories wholly devoted to the servicing of production, there were 165 trust and factory laboratories with a total employment of 3585 (equal to 15 per cent of the manpower of the industrial research institutes at the time).[25]

Continuing neglect of the laboratories in industry is implied in the on-going discussion of the position of the factories' scientific facilities with regard to the central department of the commissariat on the one hand and the branch industrial administration on the other. In July 1932 the relationship between NIS and the *ob"edineniya* and the laboratories was reviewed at a meeting in the commissariat. NIS's role was then seen as the planning of research into methods of production control, methodological guidance of R&D undertaken by factory laboratories and general supervision of production control work. The ob″edinenie's role was 'operational' control over research work and control-testing, planning and finance. The organisation of the training of laboratory workers was left to the plants themselves.[26]

An important part of these attempts to regularise the control and supervision of factory laboratories was the establishment of a central coordinating body, the Council of Factory Laboratories, which was formed under NIS in July 1932.[27] Its chairman was Bukharin, the head of NIS, and it comprised representatives of enterprises, research institutes and educational establishments. By 1936 the council had branch councils to coordinate the work of laboratories in Leningrad, the Urals and the Ukraine.[28] It was intended to fulfil NIS's general scientific supervisory functions for the laboratories, helping them solve organisational and scientific questions such as the design of laboratories, and methods of laboratory work.[29] It took over the publication *Zavodskaya Laboratoriya*, which had up to that time been published by NIS and the Committee for Chemicalisation. It was obviously hoped that the establishment of this council as a focus would help to achieve the desired improvement in the laboratories.

Notwithstanding this continuing attention there was likely to have been considerable opposition within industry to expanding the research role of the laboratories. At the same time the emphasis on developing the network of research establishments resulted in some laboratories being upgraded to form independent research institutes.[30] But the growing emphasis on R&D was fully reflected in a new set of regulations for the organisation and work of the laboratories which was approved by NKTP in September 1932.[31] The work of the laboratories was divided into three parts which were given equal weight: control work, research, and production-experimental work (development). There would appear to have been a substantial increase between 1931 and 1933 in the number of laboratories with some research capability. Bukharin in an article published in the middle of 1933 cites information which suggests, when compared with the figure from the 1931 survey, that the number of laboratories undertaking research increased by 50 per cent over the two year period (see Table 8.1). The size of individual laboratories had also grown.[32] The laboratories considered to be capable of research were most heavily concentrated in the chemical industry, but even so, as can be seen from Table 8.1, some chemical plants had no laboratories at all. There was also clearly a shortage of trained personnel especially in metallurgy. Indeed comment elsewhere suggests that these data through their superficial nature imply a greater improvement in factory science than had actually taken place, for there are reports that there were a large number of poorly-equipped laboratories which could not even cope with their factory's daily needs for control testing.[33] But the state of factory science was not now as uniformly bleak as it had been in earlier years. There were a growing number of first-class laboratories in factories such as the Svetlana works in Leningrad and the Elektrostal' steel works outside Moscow. Nevertheless such laboratories comprised only a tiny part of the total.

TABLE 8.1 Factory laboratories in 1933

Branch of industry	Number of factories	Total number	Factory laboratories Number doing research	Percentage of staff with higher education
Engineering and electrical engineering	360	222	92	19–24
Ferrous metallurgy	86	70	30	9–15*
Non-ferrous metallurgy	137	65	22	
Chemicals	173	145	almost all	24

* Frolov, *Osnovnye Zadachi*..., p. 23, gives a figure of 7–8% for the proportion of staff in the laboratories of the metallurgical industry who had higher education
SOURCE N. I. Bukharin, 'Fabrichno-Zavodskie Laboratorii – na Sluzhbu Osvoeniya Novoi Tekhniki', *ZL*, no. 7 (1933) p. 4.

A reflection of the lack of a really substantial improvement was the issuing of yet another decree in July 1933 which was aimed at improving factory science.[34] It attributed the backwardness of a lot of laboratories to two factors. Firstly there was the attitude of enterprise management. Both the heads of plants and their technical personnel were accused of failing to provide laboratories with qualified manpower and of not taking steps to ensure their satisfactory development. Secondly, some laboratories were, surprisingly, criticised for overemphasising research at the expense of work on the servicing of production. It was meant, perhaps, as a warning to laboratories not to try to run before they could walk, a reminder that good production servicing was the first priority. The decree put forward a whole series of detailed proposals for improving the position of the laboratories. Various *glavki* were told of specific enterprises to which attention should be paid.

This NKTP decree appears to have been no more successful in achieving the desired effect than those issued by VSNKh. Just as at the

Seventeenth Party Conference at the beginning of 1932 the resolution on the report of Ordzhonikidze had mentioned that the 'tasks dictate ... in particular ... *decisive strengthening of the factory laboratories* [italics in source] and the organisation of them at newly constructed plants',[35] so at the Seventeenth Party Congress two years later the resolution on the reports of Molotov and Kuibyshev emphasised the necessity for the 'broadest development of the work of scientific and technical institutes and especially of factory laboratories'.[36] There are continuing reports on the unsatisfactory state of the laboratories at meetings and conferences of factory laboratory personnel.[37] The common theme was that there had been some improvement but that there was still a long way to go. Even this level of improvement was clearly not uniform, for in March 1934 the backward state of factory science in the non-ferrous metals industry attracted special attention. The data published by Bukharin in the previous year (see Table 8.1) had suggested that the provision of scientific services in the branch's enterprises was considerably lower than in other important branches of industry; and now NKTP published a decree on 26 March which expressed general dissatisfaction with the laboratories in the branch's enterprises in the Urals and pilloried Sevtsvetmet, the *ob"edinenie* which controlled the branch, for ignoring the NKTP decree of the previous July.[38] Subsequently a special commission from the Council of Factory Laboratories was sent to the Urals[39] and five months later NKTP approved detailed recommendations for improving the work of the laboratories.[40]

The overall picture remained somewhat varied. In some plants considerable sums were now being spent on research,[41] but in some branches such as non-ferrous metals – agricultural engineering and coal were two other branches regularly cited – facilities remained poor even for control analyses and testing. A crude estimate based on data from two publications which appeared in 1934 and 1935, containing information on a selected number of factory laboratories, would seem to indicate that the research effort of the factory laboratories at this time represented the equivalent of less than 10 per cent of the resources of the industrial research establishments.[42] By the mid-1930s, notwithstanding the existence of branches where laboratory provision was poor, there seems to have been some degree of overall satisfaction with the production servicing side of the laboratories' work.[43] The major concern was now their research role and their function in the process of innovation. Considerable emphasis was placed on the relations between plant laboratories and the research institutes and on the position of enterprises' facilities within the R&D network.[44]

The emphasis on the need to build up the factory research base was also being accompanied by a parallel stress on the role of the factory in the design process, especially in the engineering industry. Indeed in 1935 NKTP's *glavki* for engineering were called on to concentrate design work directly in the enterprises, by severely limiting the functions of the central design bureau which existed in most branches of the industry to questions of overall design policy and the supervision of the work of the plant design teams. The staff which would now be surplus to requirements were to be transferred to factory design departments.[45] In contrast, at the start of the 1930s the policy had been to create centrally controlled organisations for a whole branch and existing factory resources had been utilised to form such central design bureaux – for example, the design team of the Kolomna railway engine works became the core of the design bureau Lokomotivoproekt.[46] But now there were worries about the ability of such a structure to ensure the necessary ongoing improvement in the design of the equipment and machinery which was now being successfully produced by the new plants which had been established during the previous years. The guidelines for the following year's review of the work of the industrial research network were also to envisage the transfer of such design work from those institutes engaged in design to industrial plants.[47]

Since it was proving so difficult to develop an extensive research base at the factory there was growing pressure to improve research by handing over established independent research facilities to plants.[48] The maximum development of research at enterprises was reportedly the guiding principle behind the review of industrial research in 1936.[49] Not surprisingly, during the subsequent discussions the representatives of plant laboratories were strongly in favour of any such measures which would increase the resources available for research at enterprises, stressing the importance of doing work on the improvement of existing technology in factory laboratories.[50] As we have seen above, as a result of the reorganisation in this year a small number of institutes were wound up and their resources used to improve factory scientific facilities; thus, the equipment and personnel of the Urals Institute for Non-Ferrous Metals were used to strengthen the laboratories of the plants at Krasnouralsk, Kirovgrad, Karabash and Chelyabinsk.[51]

The state of the scientific facilities of industrial plants was to continue to be a matter of concern and discussion. In 1937 a review of R&D in ferrous metallurgy resulted in a call for the establishment of laboratories doing research at all large steel plants,[52] and three years later there was still dissatisfaction with the factory laboratories in this branch.[53] In 1938 Novikov, chairman of NKTP's Technical Council, again raised the question of improving the facilities of plants with the equipment and personnel from institutes.[54]

While there was no massive injection of resources into factory research, a steady growth in factory R&D did take place. In 1938 the research expenditure in the laboratories of NKTP's enterprises was, at forty million rubles,[55] the equivalent of over 15 per cent of the expenditure on the commissariat's research institutes; this was a clear increase over our estimated funding for the middle of the decade. It would seem that such research expenditures comprised 25 per cent of the total expenditure on the laboratories.[56] Nevertheless at the end of the 1930s it was still true to say that in Soviet industrial plants R&D activities were the exception rather than the rule.[57] The role of the machine-tool factories in design, referred to above, is the one example that has been found of any systematic involvement in R&D in a branch of industry. In addition to a growing number of individual plants from different branches of industry were picked out as having a well-developed R&D capability; these included the aforementioned Svetlana electrical-engineering plant – closely involved in defence work such as the early Soviet work on radio location – and the Khar'kov Tractor factory which was also designing and building tanks at the end of the 1930s. It would indeed seem likely that there was a correlation between good factory laboratories and plants in priority fields.

The Khar'kov Tractor Factory was also an exception to the 'law' that the neglect of laboratories was worst in the newly constructed plants. Academician Fersman, in an article on the scientific servicing of the new giant enterprises, drew attention to a photograph which he had seen in an American journal: 'in it was a field, in this field was a single building with the sign "Chemical Laboratory" and the caption under the photograph was "Site of a future cement factory".'[58] He strongly contrasted this with the usual position in new Soviet factories, where the construction of laboratories came a long way down the list of priorities. In some cases they were not even envisaged at the project stage. This led one author to remark when writing about a plant (Zaporozhstal') where laboratory facilities were constructed that they were 'not only projected but even built'.[59] A particularly well-documented case of the neglect of laboratory facilities is the giant new metallurgical combine at Magnitogorsk.[60]

Thus, the overwhelming concentration on independent R&D facilities divorced from the factory was not the result of a conscious policy decision on the part of the Soviet government. While the independent research institutes which had been established after the revolution were generally looked on as the bodies which would provide major technical developments, great emphasis came to be laid on the factory in improving existing technology and from the beginning of the 1930s the development of factory research facilities became a matter of increasing central attention. However, while there was some growth in factory research, the desired large-scale expansion did not take place and at the outbreak of the Second World War expenditure on R&D within industrial plants can only have been a small part of the total expenditure on industrial R&D, and only a tiny part of the value of industrial production. Clearly an important question concerning the development of Soviet R&D is why the reviews, reports, decrees and resolutions of the 1920s and 1930s did not result in the widespread establishment of well-endowed R&D facilities in industrial plants.

The development of factories' scientific facilities, particularly in the early part of the period under review, was hindered by the lack of the necessary resources of manpower and equipment. For example, the report of the 1930 survey of laboratories and the NKTP decree of July 1933 and the discussion surrounding the review of industrial research in 1936[61] all refer to problems in the supply of suitably trained manpower. The attempt at the end of the 1920s to increase rapidly the number and size of laboratories at a time when there was a growing shortage of all kinds of technical and scientific personnel put great pressure on the possible sources of staff and led to an emphasis on factory training and the upgrading of factory personnel to responsible laboratory positions. In 1930 it was stated that 2500 laboratory workers were needed in the metallurgical industry alone.[62] Additional problems were created by the increasing emphasis on the research functions of the laboratories. Such work demanded a higher standard of personnel than analytical and testing work. There are frequent references to laboratories with only one or two highly qualified workers. The NKTP decree of July 1933 informed the Personnel Sector of the commissariat that it was to take measures to increase the percentage of workers with higher education in the laboratories of leading enterprises.[63] The poor quality of the people doing research in factory laboratories is reflected in an article published in 1935, in which the author states that 75 per cent of the reports on research done in plant laboratories contained elementary mistakes.[64] While in the second half of the 1930s there continued to be critical references to the manpower of the scientific facilities of individual factories,[65] the absence of any detailed central attention to the supply of laboratory personnel suggests that, with the now greatly increased supply of trained scientific and technical personnel, the demand for laboratory workers could by and large be met.

Laboratory equipment and materials were always in short supply. Throughout the 1930s there are references to the laboratories' inability to obtain the necessary materials to carry out their work – and also of the low quality of those that were available – and reports of proposals intended to improve supply.[66] Further, unlike many research institutes, factory laboratories did not have the necessary skills or facilities to fill the gaps in domestic production by manufacturing their own apparatus and materials. The shortages of such resources jeopardised the laboratories' production control functions, to say nothing of disrupting any research programmes.

While the importance of countrywide shortages in explaining the poor supply of manpower and equipment for factory laboratories cannot be denied, it was also a result of the attitude to the laboratories taken by the managers of industrial enterprises. For example, a handicap to laboratories' staff recruitment attempts seems to have been managements' frequent failure to respond to demands to eliminate the differential between the rates of pay earned by laboratory workers and those which could be earned by the same person in a production shop.[67] Similarly the construction of purpose-built laboratory premises or their improvement often came low on their list of priorities and earmarked funds were diverted for other purposes.[68]

Indeed the main reason why the development of the laboratories did not match the expectations apparently held for them can be found in the attitude of plant managers and their superiors in the trust *ob"edinenie* or *glavk*. In a lot of enterprises the laboratories were not looked on as having any role to play in the work of the plant, and regarded as nothing but a financial burden. They were 'situated literally in backyards, in the position of being the enterprises' unwanted children'.[69] An extreme case of the attitude taken concerns the *ob"edinenie* for non-ferrous metals, Tsvetmetzoloto, which in 1930 reportedly closed at twenty-four hours' notice the central scientific laboratory of the Mining Chemical Trust, which is stated to have been well-equipped, and to have a staff of fifty, and was headed by a foreign specialist. The laboratory premises were apparently needed by the *ob"edinenie* to accommodate its tariffs and rates (*tarifo-ekonomicheskii*) department.[70] There were also cases of 'laboratory' workers who were neither working in laboratories nor according to their speciality. It is said to have been 'well-known' that eleven laboratory workers of the steel plant Zaporozhstal' were working as clerks in the factory administration in the early 1930s.[71]

In the 1920s such attitudes could have been the result of ignorance of the value of the scientific servicing of production. By the start of the 1930s they were more likely to reflect the priorities held by plant managers as a consequence of the crude quantitative emphasis of their production plan. This form of managerial success criterion would have had a twofold effect on the attitude towards scientific laboratories. As a consequence of a resulting unwillingness to spend resources on non-productive activity, laboratories were likely to be starved of funds for research as opposed to testing and control. Indeed in spite of the emphasis placed on research from the beginning of the 1930s it was reported at the beginning of 1935 that managers still saw them as purely engaged in the control of production[72] and there is little evidence of a reversal of this view in the second half of the decade. In fact, the likelihood of tension within a plant between research and routine testing was suggested by western authors as one of the factors likely to hinder research in plant scientific laboratories.[73] On the other hand the need for the Soviet manager to fulfil the plan at all costs, in the last resort with substandard products, will have undermined even laboratories' control functions.

The attitude of the plant's superiors to the laboratories is reflected in the failure of such organs to ensure the plant's compliance with the many directives and resolutions on factory scientific laboratories. In 1940 the Commissariat of Ferrous Metallurgy was specifically criticised for not checking whether the measures it had approved in a decree of the previous year had in fact been implemented.[74]

9 Industrial Research and Innovation

A consequence of the growth of a network of independent research establishments and the failure to develop widespread R&D at the enterprises was that links between the research laboratory and the plant became a matter of much concern. Firstly, it was necessary to ensure that the work done by these independent organisations was relevant to the needs of industry. Action taken towards this end included the policy of fostering research in the areas of industrial growth and the subordination in 1929 of the majority of the research institutes to organs directly responsible for the individual branches of industry; further it was an important aim of the research planning system. Secondly, the links between research and industry had to ensure the smooth and rapid utilisation in industrial production of the results of the research network's activities. In this chapter we are specifically concerned with this second aspect – the industrial application of the research undertaken in the Soviet industrial research establishments. However, in practice these two problems were closely related in industrial science policy. Measures directed towards establishing an 'atmosphere' more conducive to close relations between research institutes and industry would not only increase the likelihood that research programmes would be of use to industry but also that the results would be taken up. Indeed initially the discussion of the utilisation of research and measures taken to try to improve it were part of such a general emphasis on the linking of science and industry. But in the middle of the 1930s innovation came to be seen largely as an independent problem. It was now realised, as Academician Fersman wrote in 1936, that the industrial application of research results was 'a new creative act'.[1]

Discussion of the innovation process and dissatisfaction with the scale of utilisation of the results achieved in the research laboratory has a long history in the Soviet Union. As early as the mid-1920s there was a feeling that the work of the industrial research institutes was not being fully utilised. The existence of research suitable for industrial application which had not been taken up was one of the factors underlying the debate about NTO VSNKh at the end of 1924 and the beginning of 1925. Dzerzhinskii, in his remarks at the meeting of the Presidium of VSNKh which discussed NTO, briefly referred to a backlog of completed research.[2] Just over two months later at the Fourteenth Party Conference he forcefully expanded these earlier comments:

> If you were familiar with the position of the technical sciences here in Russia, you would be amazed at our successes in this field. But, unfortunately, who reads scientists' works? – we do not; who publishes them? – we do not. But they are published and used by the English, French and Germans, who back and use the scientific work which we do not know how to use. These people take great pains to get the largest possible benefit from our science.[3]

Dzerzhinskii's view, therefore, was that information on research was not being made available to industry; thus, a first step to improve the application of research in industry would be to increase the flow of information. As a result, the improvement of innovation was seen as a natural corollary of any general improvement in the links between the research institutes and industry. VSNKh sought to improve these links by establishing the Councils of Assistance and the Technical Conferences to bring together research workers and industrial personnel. A function of the former was in fact to keep a watching brief over the transfer to industry of products and processes developed by institutes.[4]

This view of innovation was to persist for several years. The shift in emphasis in the work of VSNKh's central scientific organ towards a more positive role in the technical reconstruction of industry, which was first envisaged in the VSNKh decree of December 1926 on the work of the scientific and technical establishments, appears to have brought no immediate change in the emphasis on the linking of research and industry through the Councils of Assistance and the Technical Conferences. Although Sverdlov, in his report on the work of NTU which he made to the Presidium of VSNKh on becoming its head, stated that the quickest utilisation of the scientific and technical experience 'accumulated in NTU' was of especially great importance, the measures which he put forward and which were approved were essentially a reiteration of those of two years earlier.[5]

However, continuing dissatisfaction with the utilisation of research was a feature of the report of NKRKI to Sovnarkom on its study of the industrial research network in 1928.[6] The Karpov Chemical Institute was cited as an example; of eleven important pieces of research completed by the institute only three had been taken up by industry. Only one VSNKh institute, the Scientific Chemical Pharmaceutical Institute, was considered to have 'more or less satisfactory' links with industry. It considered that simple information exchange was not enough to ensure the best use of the growing amount of research. In fact the very language used in the report would appear to have reflected NKRKI's view of the existing state of affairs, for its report seems to have been the first occasion on which the term *vnedrenie* (introduction) was used to describe the industrial application of research results. This word, which was in future to be widely applied to describe innovation, implies the necessity to push against obstacles.[7] NKRKI clearly felt that a more formal structure for innovation had to be established, for it suggested that a particular body should take overall responsibility for the industrial application of research, proposing that the *glavk* should in future be accountable for any failure of its branch to make use of relevant research. The assignment of this responsibility to the *glavki* was in line with their growing importance in the future technical development of Soviet industry. They administered the project organisations which were being set up to produce plans for new factories and were shortly to be made responsible for the NTSy.

The resolution of Sovnarkom on this report[8] and the subsequent VSNKh decree[9] included the first major statement on the innovation process. VSNKh dealt in detail both with the introduction of research which had already been completed and with the path to be followed in future. As in the NKRKI report the main onus was placed on the *glavk*. NTU's function was to ensure that the *glavki* received the necessary information from its research institutes. As part of their role in ensuring innovation the *glavki* were to provide development facilities where they were lacking at institutes, by making provision (with trusts) for the construction of pilot-plants and prototypes in their subordinate enterprises; the necessary funds were to be included in the enterprises' financial plans. They were even, where appropriate, to finance the establishment of experimental installations at institutes. To provide some of the vital flexibility a sum equivalent to 10 per cent of the funds for research in the financial plans of trusts and enterprises was to be set aside for financing development not envisaged in the plan.

The VSNKh decree, therefore, foresaw that the development and testing of the results of research undertaken by the industrial research institutes would be done by the institute and/or the industrial enterprise. In the following year an alternative solution to the problem of innovation was suggested by G. P. Brailo. While he echoed NKRKI's criticisms of the poor performance in innovation, he considered that for industry itself to undertake the necessary work on development and innovation was made difficult 'not only by the limited material resources for swift technical re-equipment, but also to an important degree by traditional operating methods'; on the other hand for research institutes to undertake this work 'would mean loading them with operational and production work at the expense of research'. He concluded that the realisation of the achievements of the institutes on the requisite scale required a third type of organisation, 'independent enterprises, technical offices, project bureaux, perhaps special jointly owned companies'.[10]

Brailo appears to have been alone among scientists at this time in considering the problems of innovation. There seems to have been little discussion at the First All-Union Conference for the Planning of Scientific Research beyond a report by Brailo himself, now vice-chairman of NIS VSNKh, on the introduction of research into industry, in which he spoke of the need to eliminate as quickly as possible the gap between scientific creativity and the utilisation of its results. The majority of the proposals which he made to close this gap were the same as the measures previously suggested – institutes should be provided with experimental factories and installations, and also design bureaux; their boards should include industrial managers. However, he proved to be almost twenty years ahead of his time by suggesting that there should be overall planning of innovation on the national level – the first annual plan for new technology was approved in 1949;[11] for he pointed out that innovation on the basis of the research of the institutes was not reflected in the plans for the development of industry and suggested that it should be undertaken in a planned and systematic way 'organically linked with the plans for the branches of industry and the regions of the country'.[12]

Brailo's remarks on the need for experimental facilities clearly suggest a failure of the 1928 decree's measures to provide them; and there were, indeed, frequent references to a shortage of facilities for building prototypes or pilot-plants. The directives issued by NIS for the drawing up of the research establishments' control figures for 1930/31 picked out as a vital fault the lack of such facilities and suggested that the *ob"edineniya* were to take factories doing such work from the production network and provide them with the necessary extra equipment.[13] The guidelines for the following year proposed that the *ob"edineniya* 'in every possible way speed up' the provision of institutes with development installations, for NIS considered that the lack of such facilities was the basic reason for the poor industrial application of the results of research.[14]

In these years there would, however, appear to have been some improvement in the number of research projects taken up by industry, but the view was that with the rapid expansion in the industrial research effort the numbers of potential innovations had grown even more rapidly. E. P. Frolov, Brailo's successor as deputy head of NIS, referred to a growing disproportion between research and innovation in the early part of the 1930s.[15] He was also unwilling to attribute this solely to the shortage of development facilities, for he pointed out that innovation was often an extremely lengthy process even where the relevant institute had the necessary facilities.[16] A similar conclusion would appear to have been reached also in a study of the research network which formed the basis of a TsKK-NKRKI resolution of September 1931, which placed great emphasis on the need for institutes to play a more positive role.[17]

In spite of the growing indications of a continuing failure to make full use of the results of industrial research, the Second All-Union Conference for the Planning of Research in Heavy Industry seems, like the initial research planning conference, to have lacked any detailed discussion of the problem of *vnedrenie*. Bukharin in his opening report on the general position of industrial research only referred to it briefly in the context of the general linking of science and the factory, but he did suggest that particular factories should be attached in an informal way to particular branch institutes.[18] The rationale behind this proposal was illustrated by the conference's resolution on his report, which picked out as one of the basic faults of the industrial R&D system the insufficient links between research and production, which were 'particularly reflected in the lack of any firm structure of responsibility for the industrial application of completed research'.[19] Bukharin was proposing 'official' informal links between institutes and factories as an attempt to remedy the lack of any formal structure for innovation. This situation was basically a consequence of the changes in the administration of Soviet industry over the preceding years. As we have seen the VSNKh decree of August 1928 made the *glavk* responsible for any failure to introduce the results of domestic research into industry. However, at the end of the following year the major reorganisation of the administration of VSNKh industry took place which saw the abolition of the *glavki* and the formation of the *ob"edineniya* as the organs responsible for the individual branches of industry. They did not however retain the *glavki's* responsibility for innovation; this was now to be the task of the trusts—now with much narrower functions than before and mainly concerned with the technical direction of industry. Since the *ob"edinenie* was responsible for the overall technical level of its branch, the position of the trusts was somewhat anomalous, and it is not surprising to find that from the start of 1931 responsibility for the industrial application of the work of the research establishments was to be part of the function of the *ob"edinenie*.[20] However, the division of the *ob"edineniya* into more specialised units was accelerating, the number of independent trusts was to grow, and within months the *glavki* were to start to return. The remarks made at the conference in December 1932 suggest that no decision had been taken on the responsibilities for the introduction of research within this greatly changed administrative structure.

While neither Bukharin's report nor the resolution on it discussed the allocation of responsibility for innovation, this problem comprised a major part of the resolution of the Board of NKTP which resulted from its review of the conference's activities.[21] Indeed the brief general discussion of innovation at the conference can be contrasted with growing government concern, for this resolution was almost entirely devoted to the problem of *vnedrenie*. As in 1928, proposals were made both to secure the introduction of already completed research – some thirty examples of new products and processes were listed in the resolution itself – and for procedures to be followed in the future. These were set out in some detail and the role of each body in the innovation process specified. Thus the institutes, after checking the completed work (where necessary at the pilot plant level), had to send the technical report to the relevant trust or factory – where a plant was directly under a *glavk* – and notify the *glavk* concerned and NIS; the report was to state the proposed procedure for introducing the project into industry and name the brigade which was being assigned to the work. The industrial organisation was obliged to review the institute's proposal within a month and either take it up (with the participation of the institute's brigade) or give detailed reasons for rejecting it; it was to inform NIS of its final decision. The *glavk*, in the person of the head of its production-technical sector, was to bear the responsibility for innovation, with NIS generally supervising the process. NKTP also considered that future research contracts should include a special paragraph on the procedure for using the results of the work. Technical reports on research projects were also to be sent to the project organisations which designed new plants.

NKTP's detailed resolution, therefore, reinstated the *glavk* as the body responsible for ensuring that industry took up and used domestic research. It also reflected the recommendations of TsKK-NKRKI and the resolution on the planning conference by assigning a much more active role to the research institutes in the innovation process; this clearly presented a sharp contrast to the basically passive role which had been seen for them in the discussions of the mid-1920s, when the onus was largely put on industry to find out what products and processes had been developed. The tenor of the decree was also that the institute was to be wholly responsible for technical development.

While the Board of NKTP laid down this structure for innovation it made no provision for its efficient operation. The resolution presented no sanctions or inducements to ensure that each body fulfilled its role; and although there was no general discussion of the success or failure of these measures, the available evidence suggests that no radical improvement in the rate of *vnedrenie* resulted. For example, in 1934 a review of the role of science in the optical industry referred to the failure to take up the State Optical Institute's work, pointing out that of the thirty-seven projects done by the institute in 1933 at the request of the optico-mechanical industry only seven had been utilised.[22] Similarly an NKTP study of the plastics industry in Leningrad reached the conclusion that this branch was ignoring many of the research projects successfully completed by the Leningrad Plastics Institute and was extremely slow to introduce the few which it did decide to adopt.[23] Furthermore there was a suggestion that innovation was sometimes 'on paper' only. In an article on the cement industry it was stated that of twenty-five research projects completed by the Cement Research Institute which were generally held to have been introduced into industry, only two could be considered to have been fully mastered in the production conditions of a factory.[24] In several cases it needed the direct intervention of Ordzhonikidze himself to achieve any progress.[25]

The review of the state of industrial research in 1936[26] not surprisingly included discussion on ways to improve innovation. However, no radically new measures were suggested. The impression is given that it was believed that the existing system ought to work and that it was the fault of the bodies involved rather than of the system itself that innovation remained a problem area. The major criticism fell on the *glavki*, who in addition to being deemed responsible for innovation now controlled the majority of the industrial R&D facilities. It was felt that they were not giving the necessary attention to science. It was stressed that heads of *glavki* personally were to be responsible for the direction and control of research and innovation in the branch. As in the previous decrees of 1928 and 1933 Ordzhonikidze's decree of October 1936 instructed *glavki* to review research completed but not yet utilised. Each was to produce a plan for applying research results in its branch by the beginning of the following month. The decree did, however, seek to increase incentives for scientists to push through innovation by proposing that institutes should receive bonuses for innovation; in future enterprises were to pay institutes 20 per cent of the annual saving realised from an innovation in the first three years after its introduction. Such funds comprised a fund at the disposal of the institute's director similar to the incentive fund of the heads of enterprises; it could be used for personal bonuses for staff, or for the provision of housing and social and cultural facilities.[27] This was the first attempt to introduce a systematic incentive system. Bonuses had previously been paid both to individuals and for distribution by the institute as a reward for what was considered to be an important innovation by the commissariat.[28]

The decree of October 1936 was no more successful than its predecessors in providing a smooth transmission belt between the research laboratory and the factory floor. Two years later Kaganovich, the head of NKTP at the time, pointed out that the *glavki* had failed to incorporate into their plans the introduction of as yet unused developments, as had been demanded in the decree.[29] Complaints and editorials on the failure to make full use of available research continued to be published.[30] Special decrees had to be issued on the innovation of particular new products and processes and even these did not always result in the desired action.[31] Where innovation did take place it was an over-lengthy process.[32]

Thus, in spite of growing attention to innovation as a problem which had to be solved to enable the Soviet Union to acquire the full return from its industrial research effort, it would seem that throughout the inter-war years full use was not made of the work of the industrial research network and that potentially new products and processes were being lost along the road from research laboratory to production shop. The overall rate of technical development was poor.

The preceding discussion of this continuing analysis of the process of innovation and of measures taken in an attempt to improve the situation has referred in passing to factors, such as the lack of development facilities, which were influential in retarding innovation. We shall now consider these in more detail. The structure of the R&D network, the level of provision of the necessary facilities, the attitudes taken by the organisations involved and their responses to the environment in which they operated were all in varying degrees barriers to successful innovation.

FACTORS AFFECTING INNOVATION

Administrative and geographical barriers between research institutes and the enterprise.

The concern over the fact that the industrial research institutes were administratively divorced fom the actual process of industrial production had resulted in the subordination of the majority of them firstly to the *ob"edineniya* at the end of 1929 and later to the *glavki*. However, in spite of this, in the 1930s, they remained far less closely linked to industry than the research facilities of the industrial corporations in countries such as the United States. The attachment of institutes to factories which was first proposed at the beginning of 1933 was aimed at closing this gap between the institute and the enterprise. Later, of course, some institutes' personnel and resources were incorporated into factory R&D facilities. But in the absence of a large development of enterprise R&D, a disadvantage stemming from the establishment of research institutes and later the other independent R&D organisations was that they were necessarily somewhat distant from the enterprise. In the strictly hierarchical administrative system of Soviet industry this was a particular problem for institutes which did research of potential interest to plants under more than one *glavk*. In the 1970s the formal linking of institutes and factories through the establishment of special scientific-production *ob"edineniya* has been seen as a development which can break down this administrative barrier to innovation. In fact a similar approach was attempted on at least one occasion in the inter-war years, for at the start of the 1930s VSNKh's Central Radio Laboratory in Leningrad and the Komintern Radio Factory were amalgamated for a period of about eighteen months, but this experiment apparently did not prove to be a successful linking of science and production.[33] In the period under review the links between institute and enterprise remained tenuous and coordination of the innovation process in the Soviet industrial system complex; this was particularly true of the first half of the 1930s when, as we have seen, the research institute, enterprise, trust, *glavk* or *ob"edinenie* and NIS were all involved.

The effects of administrative separation can be offset by geographical proximity. Where research facilities are not closely linked administratively, the closer they are geographically the more likely they are to build up the necessary contact for successful innovation – a factory in Chelyabinsk is more likely to establish the necessary close links to ensure smooth innovation with an institute in Chelyabinsk than with an institute in Moscow.[34] Thus, the continuing concentration of research institutes and also of other R&D organisations in Moscow and Leningrad undoubtedly acted as a drawback to the utilisation of research. In the early 1930s Frolov of NIS, when discussing *vnedrenie*, called this concentration of institutes a 'misfortune';[35] and in an article pressing for more importance to be attached to the provincial research institutes a chemist from Sverdlovsk did indeed state that experience had shown that success in innovation declined as the distance between the institute and the factory increased.[36]

The provision of development facilities at industrial research institutes
As early as 1925 NTO VSNKh had recognised the importance of development facilities. In a book published in that year Flakserman reported that the necessity for industry to be presented with the results of research in a form directly applicable in production had led NTO to devote special attention to the checking of these results at the pilot-plant scale.[37] He went on to state that nearly all NTO's institutes had experimental factories to do this work. However, while most of the institutes might have had small installations, Flakserman's remark would seem to be an exaggeration – there were, for example, no facilities for development at the Karpov Chemical Institute[38] or at the Research Institute for Fertilisers[39] at this time.

It seems likely that in the years 1925 to 1928 the rising amount of capital expenditure on the VSNKh research institutes[40] provided a growing proportion with development facilities in the form of experimental factories or installations for testing on pilot-plant scale. The construction of an experimental factory at the Karpov institute was undertaken in these years. However, in an article published at the start of 1929, Lapirov-Skoblo still spoke of the lack of pilot-plant installations in a number of institutes, due to shortage of funds. This, he said, resulted in insufficient testing of product designs and processes.[41] As we have seen, the measures outlined by Sovnarkom and VSNKh in 1928 aimed at improving the availability of facilities by getting *glavki* to finance installations at institutes. The five year plans which the NTU institutes were drawing up at this time foresaw an expansion of development facilities; seven institutes out of twenty-three planned either to build new or to enlarge existing premises.[42]

The years after 1928 were marked by the rapid increase in the size of the industrial research network, and although the existence of development facilities was becoming more common at the old-established institutes, the total level of investment in such facilities could in no way match the growth in the number of institutes. It was reported in 1933 that the majority of the institutes in the engineering industry, including the main institute, TsNIIMash, the Central Research Institute for Engineering, did not have any experimental base, and that only in electrical engineering was there any real possibility for institutes to undertake development work.[43] In the handbook on the research institutes of NKTP which was published in 1935,[44] only twenty-one research establishments, out of ninety-eight institutes and twenty-four branches, were reported as having experimental factories, while another thirty had some development facility. As might, perhaps, be expected, such facilities appear to have been greater in institutes serving the chemical industry than in other fields. The failure to provide for any large expansion in development facilities in the years of hectic growth after 1928 is shown by the fact that the experimental factories of at least eleven of the fifteen institutes which had them dated from before the start of the First Five Year Plan.

It would have been expected that with the continuing rationalisation of the research network and the growing investment in facilities as the thirties progressed the provision of institutes with pilot-plants and other development facilities would have increased. However, the complaints about the lack of such installations in particular branches of industry and at particular establishments continued.[45]

Further, the existence of development facilities at institutes did not mean that in all cases they were in fact used for development work. There are many references to the use of experimental factories and pilot-plant installations for other purposes.[46] They were often engaged in providing the institutes with scientific equipment or materials and also supplying such products to other organisations.[47] Other institutes were engaged in straightforward manufacturing. In the mid-1930s the experimental installation of the Research Institute of Organic Semi-Products and Dyes was producing materials which had previously been imported,[48] while 99 per cent of the work of TsNIIMash's quite recently opened experimental factory was, according to the institute's director, commercial production which had been ordered from it by organs such as NKTP.[49] This production probably fell into two categories: firstly, pilot-plants were established by institutes to develop a particular product and then continued into full-scale production because the industry concerned couldn't or wouldn't take on full-scale production;[50] or, secondly, as in the case of TsNIIMash, the commissariat or *glavk* used the skills of the institute's development facilities to produce one-off models or small batches of specialised equipment. Whatever the reason, such activities clearly disrupted development work on potential new products and processes originating within the research institute. Such practices were forbidden by the decree of October 1936,[51] but with what success it has proved impossible to ascertain.

Other development facilities
Very little information is available on experimental production facilities outside the research institute network, but there were probably three other sources of development work.

Firstly some of the independent design organisations probably had the capability to build prototypes or pilot-plants. For example, the independent aircraft design organisations appear to have controlled workshops or factories where prototypes were built.[52]

Secondly, some independent experimental factories were established either to promote central prototype or pilot plant construction for a particular branch of industry – plastics and machine tools were two examples[53] – or to undertake development work in fields to which the state attached particular importance. Of the eighteen experimental factories, four experimental mines and two experimental installations listed as independent organisations under NKTP in 1935,[54] the four mines were under Podzemgaz, the body responsible for the projecting and development of experimental mines for the underground gasification of coal, two factories were for the pilot-plant production of synthetic rubber, three were subordinate to Soyuztverdosplav, the All-Union Technical Office for Hard Alloys. Of these twenty-four organisations some, such as the experimental factories for the production of nickel and tin, were involved in the starting up of production of materials which were new to Soviet industry, although widely produced elsewhere; consequently they may have been more involved in the choice and testing of foreign technology than in experimental

126   Science and Industrialisation in the USSR

development. Further, as in the case of the research institutes' development facilities, some of these organisations were probably involved in production rather than R&D; the experimental factory for machine tools was producing specialist machine tools rather than prototypes of new machines.[55]

Thirdly, another potential site for development facilities was the factory itself. Certainly before the mid-1930s the number of enterprises possessing such facilities was limited. At the Seventeenth Party Conference at the beginning of 1932 Bukharin had stated that there were still no experimental shops in factory laboratories.[56] A 1934 survey of forty-six factory laboratories has references to experimental facilities in eleven factories in the chemical and engineering industries.[57] The plants covered by this survey were mainly those which were regularly commended for their attitude to science and for their willingness to provide facilities, so it would suggest that in industrial plants in general there had not been a marked improvement since Bukharin's remarks two years before. While references to the failure of factories to play a full role in development were to continue,[58] there is not sufficient evidence to assess the extent to which industrial enterprises were provided with such facilities at the end of the 1930s. In the machine-tool industry, just as the factories came to play an important role in design, so they also built prototypes of new models.[59] However, machine-tools were probably an exception, for this branch has been distinguished from the rest of engineering in the post-war period by its degree of factory involvement in R&D[60] and in 1939 Academician Chudakov, when discussing R&D in the engineering industry as a whole, considered that enterprises did not give as much attention as they ought to experimental shops.[61] Further, development facilities in factories, as those in the research institutes, were often engaged in current production. Julian Cooper in his study of the machine-tool industry notes that experimental shops were looked on as providing reserve capacity for times when plants were under pressure to fulfil their plans and that at such times work on prototypes would cease.[62] Where experimental installations had been established in new or reconstructed plants they were possibly in some cases established to assist in the commissioning of the plant and entirely turned over to ordinary production when the plant was fully operational.

Although less information is available on the level of development facilities than on other parts of the R&D system, it would seem that in general the provision of development facilities was a bottleneck, and

Industrial Research and Innovation 127

this was a view shared by authoritative scientists in the late 1930s.[63] Also, in many cases existing facilities were diverted at least part of the time to other work.

Industry and innovation
Industry's attitude to the work of the industrial research organisations was regularly criticised. In 1927 Sverdlov said that there was a wealth of material on the oppositon which the research organisation met when trying to bring about the practical application of their work.[64] In the following year Sovnarkom was to pick out the conservatism of the heads of industry as one of the two fundamental faults in industrial research; it resulted in insufficient attention being paid by industrial bodies to the work of the research establishments.[65] The resolution, of course, was to name the *glavk* as the body to bear responsibility for any future failings in the utilisation of research. In the succeeding years the *glavki* were to be regularly criticised for their attitude to research and innovation.[66] They were singled out in the decree of October 1936, subsequently attacked for failing to take the measures embodied in that decree, and editorials in the industrial press on innovation at the end of the thirties were still referring to the *glavki* as barriers to innovation.[67] The enterprise too was similarly criticised; indeed part of the criticism of the *glavk* concerned its failure to force its plants to introduce new developments. Factories were slow to build or test prototypes and pilot-plants, and did not want to start batch production of new products.[68] At the Eighteenth Party Conference held at the beginning of 1941 the resolution which was approved on the tasks of party organisations in industry and transport particularly picked out the conservatism of the heads of industrial enterprises as a barrier to innovation.[69]

Possibly the most detailed analysis of industry's attitude was presented by Bukharin in an article which was published in December 1930.[70] Bukharin gave the following five reasons for industry's failure to make full use of science. Firstly, industry was very often conservative in the sense that it preferred to operate in the known and well-tried way, even though industrial managers 'adopt one thousand and one resolutions about "Bolshevik tempi"'. Secondly, they frequently did not understand the value to industry of a particular invention or development. Thirdly, there was too great a fear of the technical risks that were involved in any innovation. Fourthly, industrial managements had only a short-sighted view of profit and their narrow time horizon did not include technical improvements, the full results of which would appear only after a lengthy period of time. Finally, there was a general coolness towards the research institutes – industry commonly considered it 'beneath its dignity' to turn to the research institutes for solutions to its problems, although they could be of great assistance.

Such attitudes to innovation and industrial research were not unique to the Soviet Union at that time. The Balfour Committee made similar accusations against British industrialists at the end of the 1920s and reported that a 'revolution' was needed in their attitude to science.[71] Nevertheless in the Soviet Union any general antipathy to undertaking innovation was reinforced by two features of the environment in which industry operated; these were the planning system and the large-scale utilisation of foreign technology and technical assistance.

The Soviet planning system and innovation
The effect on innovation in Soviet industry of the planning system as it developed during the First Five-Year Plan has been a continuing problem.[72] The result of the concentration on gross output as the main criterion of plan fulfilment was resistance to any action which might lead to a fall in the rate of production[73] and the introduction of a new product or process into an existing enterprise would clearly have resulted in such a drop in production for a period. Retooling and the installation of new equipment might even have meant a complete halt. Further, the difficulties arising in the period of assimilation of a new product would have continued to jeopardise the fulfilment of the plan; this would lead to a cut-back in the output of the new product (and, perhaps, the non-fulfilment of the assortment plan) and the concentration on the products already fully mastered in production conditions. Pressure for maximum production also gave rise to hostility to the establishment of experimental shops and installations at the enterprise,[74] and to the desire where possible to use such facilities for ordinary production, as has been pointed out by Cooper in the case of the machine-tool industry. The problems of development were thus greatly increased for institutes without the necessary facilities and in 1935 the deputy head of the All-Union Diesel Institute was to blame the 'tension' of production plans for the sufferings of his institute as a result of having to rely on factories for the manufacture of prototypes.[75]

The import of technology and the utilisation of domestic research
The responsiveness of industry to Soviet research was influenced by

the import of foreign technology. Although it has not been possible to undertake any detailed study of the relationship between technical assistance and domestic research, it would seem that the policy of importing technology on a large scale acted as a barrier to the utilisation of the R&D being undertaken in the USSR. Since industry was to a great extent engaged on using foreign technical assistance and since its technical policy was geared towards it, it was probably inevitable that the work of its own institutes should be to some extent overlooked. It was clearly much more attractive for a *glavk* or trust to import a complete technology from abroad, which it was assumed (not necessarily always correctly) would be producing at full capacity within a fraction of the time needed to organise the development and innovation of a similar product or process on the basis of Soviet research.

Further, the great emphasis placed on the superiority of western European and American technology had as its corollary the view that anything developed in the Soviet Union was inferior.[76] Rukeyser, a Canadian engineer who worked in the asbestos mining industry at the start of the 1930s, reports that the head of the Urals Asbestos Trust visited the United States and Canada and came back with the view that Canadian technology was to be 'copied in every detail no matter whether it applied to Russian deposits or not'.[77] There were warnings against putting too great an emphasis on foreign technology, but they apparently had little effect. At a conference of the chemical industry at a time when discussion of future technical policy was coming to the boil, Yulin, head of Glavkhim, expressed concern that so much hope was being put on technical help. He said that while people were dreaming about western science, material was 'gathering dust' on the shelves of the research institutes. He was supported by a speaker who had just returned from abroad and who quoted an example of this from his own experience; he reported that when he had enquired about the possibility of obtaining technical assistance for the production of a certain chemical, he was advised to read his own country's journals, where all the required information had been published.[78]

One of the major ways in which innovation takes place is through the construction of a completely new plant utilising a new process which has been developed or producing a new product. From the late 1920s as the Soviet industrialisation drive got under way there was a mammoth expansion in investment in new or radically reconstructed plants. However, the specialised projects organisations which were established in the wake of Gipromez were primarily bodies to facilitate the import and utilisation of foreign technology.[79] In the early years of the industrialisation programme there would appear to have been poor contact between them and the branch research institutes, notwithstanding that, after the decentralisation of the administration of research in 1929, the main branch research institutes and the growing number of project organisations were both controlled by the *ob"edineniya* (later *glavki*) responsible for each branch of industry. Indeed the 1933 resolution of the Board of NKTP on innovation had to state explicitly that the research institutes had to send technical reports on completed research to the corresponding project organisation 'with the aim of using research achievements in the projecting of new and the reconstruction of existing factories'.[80] Thus, here again emphasis on foreign technology acted as a barrier to the industrial application of Soviet research, though by the late 1930s, when the first great influx of foreign technology was coming to an end, some amalgamation of the two types of organisation had taken place.[81]

The industrial research scientists and innovation
The growing stress on the active role that research institutes should play in the innovation process was matched by criticism of the attitudes of their personnel. Thus, for example, Kuibyshev, in his report on the First Five Year Plan to the Sixteenth Party Congress in 1929, attacked the institutes for failing to make every effort to ensure that industry was provided with new technology;[82] and three years later Ordzhonikidze reported to the Seventeenth Party Conference that large numbers of research workers shut themselves away in their institutes and called on science to come out of 'the "monastic" walls of the institute and invade the factory'.[83]

The scientists' attitudes to the practical application of completed research projects revealed themselves in two main ways. Firstly, there was a lack of interest in undertaking work with a technological bias. Thus, Dr Martin Ruhemann, a foreign specialist at the Ukrainian Physical Technical Institute for several years in the 1930s, reports that when the institute was asked to supply staff for a newly-constructed pilot-plant for the production of synthetic ammonia, there were few volunteers.[84] Such antipathy was also reflected in the choice of research topics.[85] Secondly, research institutes' staffs frequently felt little concern for the future of their research once it had been completed in the laboratory. G. P. Brailo in his remarks to the First All-Union Conference for the Planning of Research pointed out with disapproval that the institutes, in reporting their work, considered it finished when they could say that the results had been passed on to the organisation with which the contract for the research had been made or when they had been presented for publication.[86] In 1936 Bauman accused the institutes of not caring enough about getting their work taken up by industry.[87] Linked to such remarks was the fact that at least one institute's administration was accused of giving low priority to the construction of pilot-plant and experimental facilities,[88] and as we have seen such facilities were often used for purposes other than development work.

A thread in the continuing debate on the correct administrative structure for research and in the discussion of the possible transfer of institutes to enterprises was a perceived need to make scientists pay greater attention to the application of their research. In 1933 more direct measures were adopted with the introduction by NKTP of the policy that scientists should be seconded to factories to take part in the innovation process;[89] this, however, did not apparently become general practice, for Chudakov, discussing innovation in engineering in 1939, called for brigades to be sent from institutes to factories.[90] Further, in 1936 a move was made to provide a carrot for scientists to innovate in the form of the bonus funds accruing from savings obtained by enterprises from new technology developed by the institutes.

The likelihood of such measures producing successful innovation was constrained by an undoubted shortage of personnel with the ability to supervise the innovation process; in 1933 out of 250 engineers in the Research Institute for Fertilisers only 25 were considered to have sufficient scientific and production experience to undertake such work[91] and in 1938 Malinovskii, the deputy director of Gintsvetmet, the main research institute for non-ferrous metals, complained that very often scientists in research institutes were completely ignorant of industrial practice.[92]

An underlying reason for this general attitude on the part of scientists towards development and innovation was what has generally come to be considered a 'traditional' bias among Russian and Soviet scientists, a preference for theoretical as opposed to applied research. This in itself was, perhaps, largely a result of the relative isolation of large areas of science from national economic life in Tsarist Russia.[93] However, such an ethos was greatly reinforced by the R&D system itself. The institutional barriers between the institute and the factory and the attitude of industry to development and innovation, which meant that any institute attempting to get a new product or process developed and introduced was going to face an up-hill struggle, will have greatly reinforced any tendency on the part of scientists in research institutes to prefer theory to practice, and to see papers and reports rather than products and processes as the final goal of their work.

INNOVATION AND THE AIRCRAFT INDUSTRY

However, our previous discussion has been focused on the civilian sector of industry, and its performance in initiating new products and processes developed domestically can to some extent be contrasted with the performance of the military sectors of Soviet industry. In the latter there were substantial achievements based on Soviet research, development and design, although there is considerable debate about their relative importance compared with imported military technology.[94] For example, by the German invasion large-scale production was getting under way of the KV heavy tank and the T-34 medium tank – widely regarded as the best of its type at the time – and of a new advanced generation of Soviet aircraft. There was also successful innovation in strategic materials; at the beginning of the 1930s the first plants were commissioned to produce polybutadiene, the first synthetic substitute for rubber to be produced on a large scale.[95]

Notwithstanding the secrecy surrounding the military sector enough information exists for us to assess the reasons for the relative success in technical development in this field. In particular, substantial material has been published on the development of Soviet aviation which contains passing references to the R&D system and the organisation of the process of innovation. Further, in the early 1930s the aircraft industry, at the time undergoing a process of rapid development,[96] was picked out as an example for the rest of industry to follow. In 1932 Bukharin, speaking at the Second All-Union Conference on the Planning of Research in Heavy Industry, said that the aircraft industry was one of the most 'developed and technically progressive' branches of industry; and its main research institute TsAGI was 'a huge, unique scientific workshop in an industry which, as a whole, was looked on as a single factory'.[97] In the following year Ordzhonikidze spoke with approval of its close links with industry, which were evident from its joint work with factories on the introduction of its work into production.[98] In the mid-1930s the development of new military aircraft slowed, so that in the later years of the decade the Soviet air force's planes were generally inferior to those of the Luftwaffe;[99] responsibility for this has been attributed by various writers to such factors as the overconcentration on the politically prestigious achievement of aviation records,[100] the absence of successful central direction after the death in 1933 of Baranov, head of NKTP's aviation *glavk*,[101] a continuing lack of suitable engines,[102] the inability of Soviet industry to batch-produce advanced designs,[103] and the purges[104] which saw both the disruption of aircraft design teams (soon to be reassembled under NKVD) and the execution of Tukhachevskii, who had been responsible to a large extent for pushing forward Soviet military technology. But the end of the decade saw another period of rapid development with a rapid expansion in production capacity and the introduction of new planes such as the YaK-1, MiG-3 and LaGG-3 fighters and the famous IL-2 'Shturmovik' ground-attack plane, which demonstrated the ability of the aviation sector rapidly to develop and start production of new planes. The position of R&D and innovation in the aircraft industry both highlights the reasons for the lack of domestically based innovations in civilian industry and points up the measures taken by the central authorities to avoid such a situation arising in the defence industries.

In fact, Bukharin and Ordzhonikidze in their remarks in the early 1930s both seem to have had mainly in mind the process by which new aeroplane designs were put into production: they were talking about airframes rather than aero-engines or about process innovation in the construction and assembly of aircraft. In fact, engines were to be a constant bottleneck in the industry.

The organisation of R&D in aviation
The size of the R&D network serving aviation was itself a distinguishing feature. While each branch of civil industry was served by only one or two research institutes and design bureaux and in some cases the research institutes themselves were responsible for design, in aviation theoretical and applied research were undertaken in a considerable number of establishments and there was in addition a network of independent design organisations. The original core of the R&D network in the aircraft industry was provided by TsAGI, which had been founded under NTO VSNKh in 1918. It grew rapidly from an initial staff of 30–40 and in 1927 employed over 500.[105] Besides undertaking theoretical research in the aeronautical sciences and hydrodynamics, it was involved in applied research on aviation materials and other aspects of aircraft construction and had also become the

country's largest design organisation; in 1925 its design department under Tupolev had a staff of 200.[106] In the early 1930s, at the time of the rapid expansion of the research network, four new institutes were founded wholly or partly on the work of its departments. Two of these, the Central Institute for Aero-Engine Construction and the All-Union Institute for Aviation Materials, were to specialise in particular aspects of the industry;[107] like TsAGI they came under the *ob"edinenie* for the aircraft industry and later its *glavk*, GUAP, and then after 1939 under the newly established commissariat for the industry. This marked the start of a process whereby institutes were established for particular aspects of aircraft design and production. At the outbreak of the war there were also, for example, institutes for aircraft equipment and production-engineering.[108] In 1936 Tupolev's design department, which had already developed into a semi-autonomous unit within the institute, became an independent organisation; this was shortly followed by the separation from TsAGI of the experimental airframe construction plant which had served it.[109] These changes left TsAGI as an organisation which basically undertook theoretical research in aerodynamics and which tested designs in its well-equipped wind tunnels. Indeed at the end of the decade it was to be attacked for its detachment from aircraft design.[110] In addition to these institutes which came under the control of the industrial administration for the aircraft industry, further institutes doing some aeronautical research were controlled by the air force and the civil aviation authority – initially GVF, later Aeroflot.[111] Research was also being undertaken in the higher educational establishments training air force and aviation specialists, in particular in the air force's Zhukovskii Military Air Academy and the Moscow Aviation Institute;[112] the former played an important role in the initial work in the Soviet Union on the use of steel in aircraft construction.[113]

Design

While in the mid-1920s TsAGI contained the largest design team, other design bureaux were now being developed; these were usually small units established around one main designer. The first such bureau appears to have been formed in 1925, when the trust responsible for the aviation industry set up a central design bureau (TsKB) on the basis of the technical department of the aviation factory no. 1 – the former 'Dux' plant – in Moscow. This design bureau comprised two departments: one for land planes, under N. N. Polikarpov, was initially at factory no. 1, but was apparently soon transferred to another factory (no. 25); the other, under D. P. Grigorovch, was established at the Leningrad 'Krasnyi Letchik' factory and was to specialise in seaplanes. Within a few months these departments became independent organisations.[114]

Further design bureaux were formed in the late twenties. The French designer P. E. Richard was invited by the aviation trust to bring a design team to the Soviet Union and given facilities at factory no. 28 in Moscow. In the spring of 1929 this organisation absorbed the team previously run by Grigorovich, who had been sacked for failing to produce a successful seaplane design.[115] Later in 1929 a new small design bureau was established specifically to take over from TsAGI (now pre-eminently concerned with bombers) the planned project for the I-5 fighter[116] and there also appears to have been a design bureau under the air force.[117] At the beginning of 1930, therefore, there were possibly five independent design organisations in the Soviet aircraft industry. However, reportedly as a result of general dissatisfaction with the state of aircraft design,[118] but also probably due to a preference for large centralised organisations serving whole branches of industry, a decision was taken to create one large central design bureau, TsKB, uniting all existing design teams and concentrating them at one factory which would provide the necessary prototype production facilities. Such an organisation was set up in the spring of 1930 at the Menzhinskii 'Aviarabotnik' factory (no. 39) in Moscow. In the event it did not absorb all Soviet designers. TsAGI's design team was the most important of the design resources outside this new body. Even so, TsKB had a staff of 300 by the end of 1930.[119] It was apparently administratively under the OGPU (a predecessor of the NKVD and the KGB) and Polikarpov and Grigorovich were initially under a prison regime.[120] TsKB soon proved itself to be a very unwieldy – by the autumn of 1931 its staff had grown to 500[121] – and ineffectual organisation and the expected acceleration in design and prototype construction did not materialise.[122] After an unsuccessful attempt to concentrate design still further by uniting TsKB with the design side of TsAGI in October 1931, TsKB was broken up. It was replaced by several (at least nine) independent units specialising in particular types of aircraft – for example, fighters under Polikarpov, long-range planes under P. O. Sukhoi and reconnaissance aircraft under S. A. Kocherigin.[123] In addition to administrative separation a physical dispersal of the design teams took place.[124] In future design was mainly to take place in a series of teams grouped around a chief designer. While it is not clear whether these groups were, in fact, entirely independent organisations, they certainly had a high degree of autonomy and are frequently treated as independent units in the literature. Further they were subject to a process of almost constant change as new groups were set up to tackle new projects and old teams dispersed. As we have seen, the design side of TsAGI became part of this network in 1936. In that year there were fourteen design bureaux with a staff of 1370 (it is not known whether this includes the design staff of TsAGI); by 1939 they numbered thirty with a total staff of 3166[125] – this probably includes the considerable number who were with Tupolev in 'the first circle'.[126] Further new design bureaux were established in the following year.[127]

While these design teams comprised the major force in Soviet aircraft design, design work was also being undertaken in the 1930s in the Moscow and Leningrad institutes of the civil aviation authority[128] and in the educational establishments[129]. This design work was very largely concerned with civilian aircraft. Finally, some of the aircraft production factories had their own design facilities. When Grigorovich's design organisation was transferred from Leningrad to Moscow in 1927, some of its staff remained at the 'Krasnyi Letchik' factory and by the middle of the 1930s this plant had a design bureau.[130] Design facilities also existed at factory no. 1[131] and may have been established at some of the new aircraft plants.[132] These factory design departments were probably initially set up to assist in solving the problems which arose when batch production of a prototype was started. They do not appear to have been responsible for any important new design, although further development of existing production types was undertaken.[133] Their facilities were also used by new young designers, for example Yakovlev, to design and construct their first planes, sometimes with the support of outside funds.[134]

Thus by the end of the 1930s the Soviet Union had built up a considerable R&D network serving the aircraft industry and in comparison with other branches of industry was investing a large amount of resources in this field. In contrast to the general centralisation of R&D within the branch which existed elsewhere, aircraft design, after an abortive attempt to build such a large central organisation at the start of the 1930s, was undertaken by a whole series of small flexible units.

The links between design and production
Bukharin's remarks on TsAGI in 1932 imply that the links between this institute and the industry were close. Indeed thanks to its well-equipped facilities in the years when Tupolev's design team was based

there, TsAGI was able to do a whole range of development work on the prototypes built in its own experimental factory. It was also reported that even before the completion of this work TsAGI knew in which plant batch production would take place and the plant was brought in to assist in development with factory personnel working at the institute.[135] Indeed six months before the first flight of the TsAGI-designed TB-3 bomber, the 'Tenth Anniversary of October' factory began to study the blueprints and in the course of testing alterations were made to them.[136] Subsequently, when the blueprints were handed over to the factory, brigades of engineers (and sometimes workers) went to the factory at the same time. When batch production started TsAGI regularly tested the parts being produced and its assistance to the plant could even include the batch production of parts which the factory itself was technically incapable of producing. Furthermore, in those plants which did not have the necessary laboratory facilities TsAGI did all the control testing which elsewhere would have been done by the factories themselves. Such close supervision was apparently maintained until all the faults arising in the course of batch production had been eliminated. Similarly TsAGI's offshoot for engines, TsIAM, had the facilities to build an initial batch of engines to iron out manufacturing difficulties before production was handed over to a factory, and it too subsequently maintained close links.[137]

As a result of the opening of an extremely well-equipped plant for building prototypes at TsAGI in 1932, its designers were probably better provided with aircraft construction facilities than those in the independent design bureaux at the time. Certainly the latter do not appear to have had such comprehensive facilities for producing parts and exerting a close control over production. Indeed some of the experimental construction shops at the design bureaux may have been stretched simply to produce the necessary prototypes. Those used by Grigorovich's design bureau in the early 1930s, in spite of their grand title of State Aircraft Factory no. 5, comprised a 'microscopically small' building in a Moscow back street;[138] and at the end of the 1930s Polikarpov's design team was apparently having difficulty in getting prototypes built because they were having to share facilities with other design bureaux.[139] However, no references have been found to any general shortage of facilities for building prototypes of new designs. From the beginning many design bureaux were situated on the premises of aviation plants, for example the Menzhinskii factory and factory no. 1; and assuming that production of a bureau's design was undertaken at the same factory, such physical proximity will have helped to ensure close contacts between the design team and the production shop. In other cases members of the relevant design team or even the whole team itself went to the factory where batch production was to be undertaken.[140] This widespread participation of the R&D personnel in starting up production in plants can be contrasted with its apparent rarity in industry at large in spite of its being proposed on more than one occasion. An important consequence of such close links and good development facilities was that it enabled the whole innovation process to be accelerated, with factories preparing for production before the end of testing and development work on the prototype. In the case of Tupolev's SB-2 bomber the tooling-up for batch production started even before the prototype had flown.[141]

The factors responsible for the relative success of innovation in aircraft
When we review the process for developing and producing new types of aeroplanes in the light of those factors which were suggested as having a retarding influence on innovation in industry in general, several important differences are apparent.

While the usual administrative barriers between R&D and the factory did exist in the aircraft industry, their negative effects were alleviated by geographical factors. Some compensation was provided by the physical siting of design bureaux at factories; but this of course did not include TsAGI's design team. Further, in the first half of the 1930s it would appear that Soviet aircraft production, like TsAGI and the other major design teams, was concentrated in Moscow, although the subsequent expansion saw the development of production in the provinces.

The aircraft industry was also distinguished by the level of its provision with development facilities. An important consequence was that the choice among competing designs could be made after the testing of prototypes. Thus, in response to government demands for a new single-seater fighter from Polikarpov and TsAGI, the I-5 (TsAGI and later the special I-5 design bureau), the I-6 (Polikarpov) and the I-8 (TsAGI) were designed and constructed in the years 1928 to 1930. Moreover three prototype versions were built of the I-5 and two of the I-6 – the I-6 and the I-8 were never to go into batch production.[142] Polikarpov is reported to have built two and a half prototypes of his designs – the half was for static testing.[143] In branches of industry without the experimental construction facilities of aviation any choice among competing designs will have had to be made at the end of the initial design stage. The aircraft industry's relative abundance of experimental factories or workshops – and of R&D organisations themselves – is clearly a reflection of a high priority attached to technical progress in aviation, which led to the necessary funds being more readily available.

Indeed, the great importance attached to the development of the air force and domestic aircraft production was the basic factor responsible for the different nature of the innovation process in this branch of industry. One of the manifestations of this importance was the close control which was exercised over the aircraft industry by the highest party and government organs. Thus, when in 1931 TsAGI proposed the construction of a plane to capture the world distance record, the Soviet government's highest military authority, the Revolutionary Military Council of the USSR, appointed a special commission to oversee the project, headed by Voroshilov himself.[144] At the time of the urgent programme in aviation immediately prior to the war the Politbureau obliged directors of aircraft and engine factories to report daily on output, and commissars and directors of enterprises providing equipment and materials for the industry were to be held personally responsible for completing orders on time.[145] These are but two examples of what was a constant involvement by the leadership in matters concerning the aircraft industry. In such an environment the head of the *glavk* responsible for this industry would have less freedom of manoeuvre than the heads of *glavki* in less important fields. The case of Yakovlev, the fighter designer, illustrated this; for it would have been very unlikely that such a relatively unimportant person as Yakovlev was in the field of aeroplane design in 1933 would have succeeded in obtaining an interview with the head of the Central Control Commission and in getting pressure put on *glavk* 'bureaucrats' (Yakovlev's own description) to give him his own design facilities, if he had been working in a branch of industry which was less closely watched.[146] Further, while in civilian industry heads of *glavki* were criticised for entrusting them to second-rank personnel,[147] R&D and technical progress were given an important place in the administrative hierarchy of the aircraft industry. In 1936 Tupolev himself was appointed chief engineer and first deputy head of the *glavk*,[148] a post which he presumably held until his arrest. Ilyushin was head of the *glavk* in 1938[149] and in 1940 Yakovlev was appointed deputy commissar of the recently established commissariat for the aircraft industry with special responsibility for experimental aircraft construction and science; indeed there appears to have been a *glavk* of the commissariat specifically devoted to experimental construction.[150]

There are two areas in which the results of government pressure could reasonably be thought to have led to some breaking down of barriers: the alleviation of the worst effects of the administrative structure and of the planning system. Thus the seemingly successful coordination of the work of design bureaux which were under other departments with the production facilities of GUAP suggests that this pressure was able to overcome inter-departmental barriers.[151] Similarly the close links that the factories established with the relevant design organisations well before the beginning of batch production of a new model and the active role that plants are reported to have played implies that central government used its authority to put pressure on enterprise managers (and also *glavk* heads) and thus avoided the worst consequences of the great importance attached to gross output. At least one plant was initially loath to give up production of the aeroplane it was producing and introduce a completely new design.[152]

The importance of aviation also appears to have resulted in the application of a more detailed planning system and the compilation of the equivalent of a plan for new technology; for Shavrov refers to a five year plan for experimental aircraft construction which was prepared by the aviation trust in 1927 and approved in 1928. The design of the I-5 and I-6 fighters was foreseen in this plan.[153]

A further reflection of the priority attached to aviation was a continuing emphasis on social mobilisation. Huge public interest was created in aircraft and aviation both by party and non-party organisations. In 1923 a Society of Friends of the Air Force (ODVF) was established to support the development of the air force, to organise the collection of funds for the construction of aeroplanes and generally to popularise aviation. ODVF and its successor Osoaviakhim (founded in 1927) became the main social organisation for supporting aviation. It had an air-school in Moscow with workshops where aeroplanes could be constructed.[154] It also financed the construction of designs produced by young and unestablished designers,[155] even importing engines from abroad.[156] Yakovlev was a regular recipient of Osoaviakhim's backing in the years before his final success in establishing his own design bureau.[157]

There was also a stress on the need for members of the Komsomol to take an active part in the work of fostering aviation, both through participation in Osoaviakhim and by enrolling in the aviation educational institutes, setting up aeroclubs and working in the aircraft industry. Particular attention was paid to the role of the Komsomol in aviation at the Ninth Congress of VLKSM in 1931.[158] A great burst of activity apparently resulted with the formation of many aeroclubs, gliding groups and aero-modelling organisations.[159] This social mobilisation, besides reflecting the priority attached to aviation, may also have resulted in the attraction to work in the field of a considerable proportion of the cream of the new generation of Soviet R&D personnel.

Indeed, another distinguishing feature of design and development in aircraft was the greater than average importance attributable to the individual. The emergence of design teams concentrated around particular designers and the failure of the attempt to establish a large central design organisation in themselves suggest that there was a greater role for the individual in aircraft development than in other parts of industry. The aim of a young designer was clearly to get into a position where he could work on his own ideas and not spend his time doing work for someone else. The persistent attempts of Yakovlev to get his own design facilities have become well known.[160] His was by no means a unique case and other designers experienced and overcame similar opposition when trying to get themselves independently established.[161]

Individual initiative in aircraft design was enhanced by the apparent flexibility of the planning system in aviation development. The available evidence suggests that there was considerable possibility of getting new projects added to the plan after its original approval and that sizeable reserve funds existed for financing such work. These projects could include additional designs to compete with those already requested from design organisations. For example, when the project for the I-5 fighter was reassigned to a specially established design bureau at the end of 1929, work on a fighter to meet the official specifications for the I-5 and I-6 fighters continued at TsAGI. Some of its staff produced an outline design for a competing single-seater fighter in their spare time. This was approved by the air force and included in the plan for experimental aircraft construction under the designation I-8.[162] In other cases design projects were suggested and approved for which there was no prior government demand. In 1933 I. V. Chetverikov began to design a long-distance reconnaissance seaplane of a type for which no official specifications existed. However, he was able to sell the idea to the Chief Administration for the Northern Sea Route as an aeroplane for use in the Arctic. His design bureau was transferred from GVF to GUAP NKTP and he obtained the necessary financial support and was provided with construction facilities at Sevastopol.[163]

It is clear that while the majority of industrial R&D personnel may have seen their work as complete when their designs were on paper, the aircraft designers were prepared to go to some lengths to get their designs built and that they were greatly assisted by the relative abundance of experimental construction facilities. It was a competitive situation and a possible harmful effect may have been some unwillingness for one design team to help another when it could have provided some assistance over a particular matter.[164] Aircraft designers were also more prepared to undertake an active part in the process of innovation itself. They did not look on their task as finished with the construction and successful testing of a prototype, but only with the completion of the lengthy and complicated process of turning this prototype into a batch-produced aeroplane. We suggested in our discussion of the civilian sectors of industry that the environment in which R&D personnel worked would have tended to reinforce any initial prejudice against development work or participation in the introduction of a product or process into industry. In the aircraft industry, the greater provision of the necessary facilities, the closer physical proximity to production and above all the close government control created an environment much less resistant to innovation. At the same time the outlook of the designers themselves may have been different; for both the 'old' and 'new' generations of aircraft designers appear to have had close links with aircraft production. Grigorovich had been head of an aircraft plant before the revolution[165] and Polikarpov had been in charge of the batch production of the 'Ilya Muromets' bomber;[166] Yakovlev started as a production engineer in the Menzhinskii plant[167] and Lavochkin took part in the introduction of the TB-3 bomber into batch production.[168]

Thus the reasons for the satisfaction expressed about the innovation process in the aircraft industry in the first part of the 1930s and for its ability to develop rapidly and produce new designs at the end of the decade were to a very large extent a reflection of the priority attached to this branch. As a result funds were made available to finance competing designs and to provide prototype construction facilities, and close central control resulted in a greater degree of flexibility in the administrative and planning system. Consequently design teams were to a considerable extent project-oriented and there was greater opportunity for group or individual initiatives. At the same time the R&D network was more closely linked to production and the individual designer was possibly more strongly committed to seeing his work introduced into large-scale production.

10 Conclusion: Yesterday and Today

The period 1917–40 not only covered the formative years of the Soviet economic system but also saw the establishment of the basic structure of its R&D system. In particular, the hectic years of the First Five Year Plan saw the formation of R&D organisations serving all branches of industry and by the middle of the 1930s the investment of resources in such activities was substantial even in comparison with the more industrially developed countries.

This R&D effort was, as we have seen, based on the formation and development of administratively independent organisations. The important position that the research institute came to occupy in Soviet industrial R&D was a consequence of the attitude of scientists, science policy-makers and industrial managers and also to some extent a legacy of the nature of Tsarist Russian industry. Immediately after 1917 the formation of such institutes was the result of a response of some scientists to the opportunities presented by the new Bolshevik government's commitment to science and the outcome of the scientists' ideas, expressed before 1917 by writers such as Vernadskii, on the need for specialised research institutes. The subsequent predominance of the institute was the result of three circumstances. Firstly, there was the absence of any widespread development of R&D activity at industrial plants before the revolution. Secondly, it became a policy within VSNKh to concentrate available scientific manpower in central organisations to serve whole branches of industry; this was closely linked to the role of scientists as industrial science policy-makers in NTO VSNKh and the development of the view that this form of organisation was superior to the unnecessary duplication of facilities which was the inevitable consequence of the barriers of secrecy among capitalist concerns. The continuance of this policy owed a lot to the 'rightist' views of Dzerzhinskii, who supported the concept of a central scientific department and central institutes against the 'leftist' assault led by Pyatakov, who wanted industrial research removed from the apparatus of the commissariat and vested in the industrial trusts. When at the end of the 1920s decentralisation of research became a fact, rapid growth in the industrial research network was already under way on the basis of building up independent institutes; and so this form of organisation survived to become the main plank of the R&D system. Thirdly, and linked to the survival of the institutes, there was the general failure of the attempts from the end of the 1920s to build a strong R&D base at the factory itself; in this case any existing conservatism on the part of enterprise managements was greatly reinforced by the stress placed on maximum output by the planning system. Similarly design (and to some extent development) was undertaken largely in independent organisations.

It is impossible to draw detailed conclusions about the success of the Soviet industrial research effort and of the R&D system in the various branches of industry from the general approach adopted in this study. However, the discussion of the industrial application of research, which was marked by the Soviet government's growing concern with innovation in the 1930s, clearly suggests an apparent inability of the Soviet Union to get the greatest return from its investment of resources in industrial R&D. Our review of the industrial application of domestic R&D pointed to three factors which were likely to have lowered the possible return on the effort invested in R&D. They were, respectively, the allocation of resources within the R&D sector, its organisational structure and the environment in which it operated – the effect on industrial R&D of the wider economic system of which it was itself a subsystem.

The allocation of resources within the R&D sector could have affected 'output' in three separate ways. Firstly, funds may have been misallocated between R and D. Indeed, this would appear to have been so; the continuing references to shortages of development facilities suggest that the relative shares of R and D within the total R&D budget were not optimal and that by spending more on D and less on R, the Soviet Union might have increased its return on total R&D expenditure by carrying a greater proportion of projects beyond the 'paper' stage. The example of the aircraft industry with its relatively good provision as regards prototype construction facilities would seem to support this view. Secondly, the purchases of 'inputs' into the R&D system may also not have been optimal. Here again, in this period, particularly before the middle of the 1930s, the material points to an incorrect balance between expenditure on manpower and expenditure on capital and current equipment. This was most clearly marked in the period of rapid manpower growth in the First Five Year Plan. Since manpower growth outran the provision of the necessary facilities, falling productivity could only result. Thus a redistribution of R&D expenditure from wages and salaries to facilities, equipment and materials might have increased 'output'. However, in practice, hardware remained in short supply in some cases throughout the period. Indeed R&D 'output' might have been raised by a transfer of funds from R&D to building up manufacturing capacity for the needed equipment, materials and chemicals. This was something institutes were in fact doing on an informal basis through their role in the manufacture and supply of materials and equipment from their own workshops. Finally R&D funds may have been spread too thinly among the various fields. If it is accepted as possible that each branch had some minimum critical funding level below which an R&D programme for the development and innovation of new products and processes was non-viable, then the blanket attempt in the Soviet Union to establish R&D facilities in all branches of industry may have resulted in a lower return than a concentration of resources in a few areas. Clearly detailed studies of the R&D programmes in individual branches of industry would be needed to assess whether such a diminution of resources actually occurred.

The organisational structure for R&D, with its stress on independent organisations, exhibited several barriers to the flow of technical developments to industry. The failure of the attempt to develop widely factory laboratories and design and development facilities resulted in a separation of virtually all R&D from production. Furthermore the available data suggest that a possible approach to the linking of R&D to production by means of close geographical links did not take place and that, in spite of a frequently cited policy for the geographical decentralisation of research, industrial R&D manpower remained largely concentrated in Moscow and Leningrad.

Lastly the period under review saw the inauguration of the planned industrialisation drive. The actual planning system and the industrial development programme's heavy reliance on the import of technology both had a fundamental effect on the performance of the industrial R&D system. The discussion of the Soviet Union's factory laboratory network illustrated the consequences for R&D at the plant of the introduction of a planning system in which the major success criterion was maximum output. It similarly raised barriers to innovation and aggravated the problems arising from the organisational separation of

R&D from production. Institutes found it difficult to persuade enterprises to build prototypes or to do other development work on their behalf and the potential cost of introducing a new product or process into production could be substantial, as a plant's output figures would reflect the tooling-up process and the difficulties of the initial commissioning period. However, the introduction of new products or processes into existing plants played only a minor role in the development of Soviet industry in these years. The industrialisation drive was based on the construction of new plants (or the radical reconstruction of existing factories). It was clearly beyond the ability of the Soviet Union to provide for the raising of the technological level of industry from its own resources and the result was the massive importation of foreign plant and technical know-how. This concentration on foreign technology acted as a barrier to the utilisation of the results of domestic R&D. The project organisations which were responsible for the designs of these new plants were geared to the utilisation of foreign technology and, notwithstanding the fact that both project organisations and research institutes were frequently controlled by the same organ, there was apparently little contact between them. Similarly the *glavk*, which was largely responsible for technical policy in its branch of industry, appears often either to have shown little awareness of the results of Soviet R&D or to have written off a domestically developed product or process as automatically inferior.

The environment in which the Soviet R&D system has operated in the 1960s and 1970s is clearly different. The economic base is built, the vast investment in new plants has taken place. It should no longer be a question of central government imposing advanced, mainly imported, technology from above on a backward industrial structure, but of Soviet industry itself generating ongoing technical progress in existing plants. However, technical progress has in general been poor[1] and the reasons suggested for this lack of development reflect very closely the discussions of the years before the war.

An initial and obvious reason is that there has been no radical change in the organisational structure for R&D; although the stress on the development of research-based corporations, the scientific-production *ob"edineniya*, has produced the beginnings of such a change. The massive injection of resources into R&D which took place from the mid-1950s – a doubling of budgetary expenditures on research between 1956 and 1960 and a further doubling in the next six years[2] – was into an R&D system basically the same as that laid down in the twenties and thirties.[3] The major development was the loss by the

Academy of Sciences of its position as the coordinator of applied research and of the quite considerable number of technically-oriented research institutes which it had acquired since the end of the 1930s.[4] The Academy's role of overseer of applied research went in 1961 to the newly created State Committee for the Coordination of Scientific Research, which was to be transformed into the State Committee for Science and Technology in 1965.[5] As we have suggested above there are several parallels between the functions of this new committee and those envisaged within the more limited area of industry alone for NTU VSNKh at the end of the 1920s in the debates on the administrative reform of industry. The Academy's industrially-oriented institutes were transferred to the authority of the state committees – ministries since 1965. With the R&D system much the same, the organisational barriers had remained, with R&D, as before the war, carried out in independent organisations. They were also still in most cases geographically separated from industry, notwithstanding continuing references to the need to expand research in the provinces. The most notable attempt to develop science in the new heartland has stemmed from the formation in 1957 of the Siberian Division of the Academy of Sciences in Novosibirsk.[6] The continued predominance of the independent R&D organisation was to be accompanied by a continuing stress on improving the still lagging R&D capabilities of enterprises.

The state of factory R&D reflected a continued existence in the economic system of those features which were anathema to factory research and innovation in the pre-war years. In spite of some changes in the economic mechanism and a greater stress on financial criteria in the planning system, the economic system has been basically the same as that which pushed through the great industrialisation drive of the 1930s. There has consequently been little alleviation in the systemic barriers to innovation. For example, 'non-productive' development facilities have remained in short supply, and the desire to use those which do exist for straightforward production purposes in the interest of plan fulfilment is still evident.[7] There has, therefore, been no radical change in the research worker's frequent view of innovation as a lengthy and difficult process, which it might be best to avoid.

As in the 1930s the state of the R&D system has led to discussion and debate on ways to improve its performance. These debates have in fact been an integral part of the general review of the economic mechanism and of discussions of the need to adjust the operation of the Soviet economy to the different demands of economic growth in the second half of the twentieth century. Further, like these debates on wider economic questions, reviews of ways to improve the performance of the R&D system have reflected the earlier discussions of the inter-war years.[8] Thus we find that in order to try and develop the close links between science and industry which organisational separation made difficult, the Soviet government in a decree of April 1961 reintroduced *khozraschet* for the research institutes and design organisations serving industry and some other branches of the economy. Now again, after a period of more centralised finance – perhaps a legacy of wartime centralisation – research organisations were to derive most of their income from contracts tied to specific pieces of work. The actual details of the contract system reflected quite closely those of thirty years previously. Similarly, the reports of the system in operation are reminiscent of the remarks on the operation of the customer-contractor principle in the 1930s.[9] Indeed the authors of Part V of the OECD report *Science Policy in the USSR*, writing in the late 1960s, sum up the effects of contract research as follows: 'In spite of the weaknesses in the contract system, it appears to have been of considerable importance in bringing industry and both academic and industrial research together'[10] – a conclusion which reflects this study's assessment of the effect of the introduction of *khozraschet* in the early 1930s. Moves towards a detailed contract system were also to be accompanied, as in the earlier period, by discussions of ways to improve the effectiveness of research and of measuring the return to research and producing a meaningful guide to research output.[11]

However, it is not financial or planning reform which is seen as the way to cut the Gordian Knot in the 1970s, but the reform of the organisation of R&D and industry based on the bringing together in one organisation of research institute, design organisation and production plant. These scientific-production *ob"edineniya*, the largest of which like the *ob"edineniya* of the early 1930s have a place in the industrial administrative structure equivalent to *glavki*, are expected to provide the core of future technical progress.[12] While, as we have seen, for a brief period the Central Radio Laboratory and a factory were joined together on much the same lines at the beginning of the 1930s, the widespread development of such organisations is a wholly new phenomenon. It reflects the differences between the demands on the economy in the 1930s and those of the 1970s and the need today to innovate not mainly by building completely new plants but by changing the products and processes of already existing factories. In their structure they do not mirror the demands of the 1930s to use the research institutes and design organisations to strengthen R&D at enterprises, but rather comprise the attachment of industrial plants to the R&D network – the head of the *ob"edinenie* is usually the director of the research institute around which it was formed. Their closest parallel in the 1930s was a research institute such as TsIAM, the Central Institute for Aero-Engine Construction, which as we saw in Chapter 9 was able to iron out possible manufacturing difficulties in a new engine by producing the initial batch of engines itself.[13] The formation of the new *ob"edineniya* may in fact be a civilian reflection of the way in which R&D, innovation and production have been organised in the priority military sectors of the Soviet economy.

The dichotomy that appears to have existed in the 1930s between priority areas such as aviation and the civilian branches of industry has also remained a feature of the R&D and innovation process. Just as the Soviet Union was able to produce tanks such as the T-34 and advanced aeroplanes, so in the post-war years it produces ICBMs and SAM missiles to a high technical level.[14] The reasons for this success in the military field which are cited by western observers are the same as those which were operative in the aircraft industry in the 1930s.[15] In this area development expenditures are relatively high, so that each design team has its facilities for prototype construction. The importance attached to it results in a very close supervision by the central leadership, which ensures that decisions taken with reference to a project are in fact carried out and alleviates the inter-departmental and inter-organisational barriers which bedevil the rest of Soviet industry. Central planning is comprehensive but at the same time flexible, which enables the relatively speedy adoption of new projects. The level of resources invested in these fields also results in an increased level of competition between competing ideas and designs since the funds are available to proceed to the prototype stage with several designs, as in the case of the I-5 fighter at the end of the 1920s.

The foregoing discussion suggests that there were several features of the Soviet industrial R&D effort which probably prevented the achievement of the fullest possible return on the resources spent in this field and that these have in many cases remained in the post-war years when the overall rate of technical progress in Soviet industry has given cause for concern. The discussions on the way to improve R&D and innovation and the measures introduced to this end have, in an expanded and developed form, reflected the first attempts to analyse the same problems in the 1920s and 1930s. However, to reverse the biblical phrase, we should not, perhaps, 'visit the sins of the children upon the fathers'. The failings of the inter-war years must be placed in their historical context. Although the years after the First World War saw the publication in the industrially developed countries of a growing literature on the importance of science in industry and on the role of the industrial research laboratory, large-scale discussion of 'science policy' and of the relationship between R&D and economic growth are much more recent phenomena. The inter-war discussion was of 'research', for the scope and concept of 'R&D' had not yet been established. Little information was available to show the importance of 'development'. It was in the Soviet Union that the problems experienced in the industrial application of research had brought in the 1920s and 1930s a growing appreciation of the vital role of development and of the need for substantial funding of this aspect of the R&D process. Similarly, in the Soviet Union in these years there was the first attempt to systematically 'engineer' innovation and the period was to be marked by an increasing realisation that there were many barriers along the road between the research laboratory and the production shop which could impede the flow of new products and processes into industry. The result was the beginning of the first comprehensive discussion and analysis of the R&D and innovation process; of the problems surrounding its organisation, planning and finance. In Britain, Soviet concern with the relationship of science to industry was paralleled at the end of the 1920s in the remarks of the Balfour Committee about the poor utilisation of science by British industry.[16] That its final report went no further than a plea for a revolution in industry's attitude towards science illustrated the great contrast between the capitalist countries and the Soviet Union in the role of the government in innovation. In Britain the problem of ensuring the swift and smooth utilisation of R&D was seen as essentially an internal matter for each firm.

Against the background of poorly developed 'science policy' in the West, the attention given to industrial R&D and science in general in the Soviet Union is impressive. The fact that such a relatively underdeveloped country, when undertaking its investment programme for industrial growth, was prepared at the same time to put a growing proportion of its scarce resources into R&D is in many ways more remarkable than its performance on the economic front during these years.

Appendix 1 The Data for Soviet Expenditure on R&D

The research network was financed both through direct allocations from the state budget and by 'non-budgetary' funds. The latter were funds which were allocated to it by enterprises and state organs as grants or on the basis of contracts for particular projects, or obtained in the form of income from the research organisations' activities.

1 *Budgetary expenditure on science*
From the budget of 1927/28 figures are available for expenditure under the budget category 'science' (*nauka*) or, as it was sometimes called, scientific research work (*nauchno-issledovatel'skie raboty*), scientific research establishments and measures (*nauchno-issledovatel'skie uchrezhdeniya i meropriyatiya*) or scientific establishments (*nauchnye uchrezhdeniya*). This comprised a part of the sector of the budget devoted to 'Education' which was itself a part of the major division 'Social and Cultural Measures'. For the years 1923/24 – 1926/27 we can obtain a total for budgetary expenditure by the addition of the individual items of expenditure made by the commissariats and departments on the various types of scientific establishments. The totals reached in this way would appear especially in the years 1923/24 and 1924/25 to be underestimates of total budgetary expenditure (see Table A1.3).

The budgetary category 'science', however, includes expenditures on establishments such as museums, libraries and archives not considered R&D organs by the Frascati definitions;[1] in a detailed breakdown of the 1935 budget, expenditures on such establishments comprised about 8 per cent of budgetary expenditure on 'science'.[2] Funds spent on the training of postgraduate research students at research establishments, another non-R&D activity, were also included (roughly 3 per cent of budgetary expenditure in 1935).[3] Further, the category 'science' includes the social sciences and humanities, and again funds spent on these fields are not considered to comprise part of R&D according to the Frascati definitions (they were perhaps 6 per cent of the 'science' budget in 1935).[4]

On the other hand there was expenditure through the budget on the Soviet Union's research establishments which was not included in the 'science' category. Elsewhere in the sector 'Education' some small amounts – well under ten per cent of budgetary expenditure on 'science' – were being allocated in the 1930s to 'the training of scientific personnel and scientific work in higher educational establishments'.[5] Other expenditures appeared in the 'National Economy' division of the budget; these were, in the main, capital expenditures on the research organisations under the economic commissariats.[6] In 1935 such funds comprised the equivalent of about six per cent of 'science' funding.

In the data published on the basis of the surveys of research establishments which were undertaken in the first half of the 1930s, the source of funds is simply given as the state budget and includes all budgetary funds whether from the 'science' category of the budget or elsewhere.[7]

2 Non-budgetary expenditure on science

The first opportunity for research establishments to obtain finance other than through direct budgetary allocation came with a decree issued by STO, the Council for Labour and Defence, in July 1923. It gave various VSNKh organs and establishments the right to additional funds – just over 2 million rubles in the first instance – which were to be given to certain branches of industry to distribute to the bodies named in the decree; these included NTO and its subordinate research establishments.[8] While at the outset these non-budgetary funds were not tied to individual research projects, in March 1927 steps were taken towards the finance of research on the customer-contractor principle when VSNKh as a result of a resolution of STO and Sovnarkom of the preceding January decided to include in the financial plans of industrial trusts and enterprises sums of money which had to be spent through contracts (*dogovory*) with research establishments.[9] These funds appeared as an item in the capital investment plan of industry. This was to be a most important development in the financing of research, especially for establishments under the industrial commissariats.

Tables A1.1 and A1.2 contain the data for expenditure on science and the industrial research establishments. Summary information can

Appendix 1 The Data for Soviet Expenditure on R & D

be found in Tables A1.3 and A1.4. It can be seen that the data for budgetary expenditure on 'science' comprise by far the most comprehensive series and that at the other extreme little information has been found on the funding of industrial research during the second half of the 1930s. Not all the available data have been included in the tables; for, in the case of budgetary expenditure in particular, there are years for which the various sources contain slightly different figures. In these cases a figure which is part of a series for several years is preferred. Some of the alternative data are cited in the notes to the tables.

154 Science and Industrialisation in the USSR

TABLE A1.1 Expenditure on 'science' in the USSR 1923/24–1941 (million rubles, current prices)

	1923/24	1924/25	1925/26	1926/27	1927/28	1928/29	1929/30	1931[p]	1932
I BUDGETARY EXPENDITURE									
(1) Expenditure through the budgetary category 'science'									
(i) All-Union budget	5.8[1]	5.4[4]	13.3[7]	15.7[10]	27.4[14]	45.7[20]	132.4[20]	145.7[20]	187.4[20]
(ii) Republican budgets	5.6[b2]	16.5[5]	[24.6]	30.0[e11]	38.8[15]	29.0[20]	41.6[20]	88.7[20]	102.9[20]
(iii) Local budgets[a]				6.2[e16]	5.2[20]	11.4[20]	8.9[20]	17.4[20]	
Total	—	—	37.9[8]	45.7[e12]	66.0[17]	89.9[20]	185.4[20]	243.4[20]	307.7[20]
(2) Expenditure on 'science' through other budgetary categories	—	—	—	—	8.3[h18]	13.2[hu18]	34.0[u22]	__ku	__ku
Total budgetary expenditure	—	—	—	—	—	—	—	—	—
II NON-BUDGETARY EXPENDITURE	0.6[c3]	1.4[c6]	3.2[d9]	5.2[f13]	32.1[g16]	31.1[mo21]	34.3[emo21]	__or	__o
III TOTAL EXPENDITURE ON 'SCIENCE'	—	—	—	—	116.8[j19]	164.6[j9]	306[n23]	495[23]	646[e23]
									470[33]

	1933	1934	1935	1936	1937	1938	1939	1940	1941
I BUDGETARY EXPENDITURE									
(1) Expenditure through the budgetary category 'science'									
(i) All-Union budget	213.2[824]	243.9[26]	349.7[29]	664[35]	434.9[38]	—	—	519.1[43]	—
(ii) Republican budgets	130.2[824]	[153.9][27]	212.4[29]	[132.5]	417.6[38]	} 425[e40]	—	} 615.8[43]	—
(iii) Local budgets		35.8[27]	78.0[e30]				—		—
Total	323.6[25]	[433.6]	610.0[e30]	797.5[36]	852.5[38]	1016[e41]	903[e42]	1134.9[43]	1032[e45]
(2) Expenditure on 'science' through other budgetary categories	__ku	100.0[khu34]	33.8[ku31]	__u	__ku	—	—	—	—
Total budgetary expenditure	343.4[824]	—	—	—	—	—	—	—	—
II NON-BUDGETARY EXPENDITURE	370.4[g024]	413.5[g028]	415[o32]	__oq	242.6[os39]	—	—	—	619[e45]
III TOTAL EXPENDITURE ON 'SCIENCE'	713.8[824]	813.1[g28]	—	1000[37]	—	—	—	3000[44]	—

The figures enclosed by [ ] are additions or subtractions of other figures in this table

See p. 155 for notes.

Appendix 1   The Data for Soviet Expenditure on R & D

NOTES TO TABLE A1.1

a   Before 1928 the autonomous republics were financed through the republican budgets, see R. W. Davies, *The Development of the Soviet Budgetary System* (Cambridge, 1958) p. 298.
b   Expenditure through the RSFSR budget only.
c   Expenditure on the establishments of NTO VSNKh, which would seem to be the only research establishments receiving non-budgetary funds at this time.
d   Expenditure on various research establishments of VSNKh, TsIK and Sovnarkom.
e   Planned figure.
f   Expenditure on the research establishments of VSNKh (actual) and on the Academy of Sciences (planned) only.
g   Survey data.
h   Expenditure on the establishments of NTU VSNKh only.
j   Soviet estimate.
k   The reports on the 1931, 1932, 1933, 1934, 1935 and 1937 budgets all include figures for expenditure on the training of cadres and research work at educational establishments; these grew from 7.8 million rubles (all-union establishments only), *Otchet Narodnogo Komissariata Finansov Soyuza SSR ob Ispolnenii Edinogo Gosudarstvennogo Byudzheta Soyuza Sovetskikh Sotsialisticheskikh Respublik za 1931g.* [Leningrad, 1932] pp. 98–100) to 52 million in 1937 (*Otchet ob Ispolnenii Gosudarstvennogo Byudzheta Soyuza SSR za 1937 God* [Moscow, 1938] pp. 58–115, 130–71).
l   Capital expenditure funds.
m   Funds to establishments under all-union organs only.
n   Figure for 1930.
o   K. Plotnikov, *Byudzhet Sotsialisticheskogo Gosudarstva* (Moscow, 1948) p. 334, states that the total amount of non-budgetary funds was 613 million rubles during the first plan and 2040 in the second.
p   In 1930 a special budget was drawn up for the period October to December to change the budgetary year to coincide with the calendar year; no information on it has been found.
q   In his report on the proposed budget for 1936, Grin'ko, the commissar of finance, stated that budgetary funds (818 million rubles) would comprise only about half of the total expenditure on science, including research at enterprises: G. F. Grin'ko, *Financial Program of the USSR for 1936* (Moscow, 1936) pp. 47, 49.
r   *Kul'turnoe Stroitel'stvo SSSR v Tsifrakh: ot VI k VII S"ezdu Sovetov (1930–1934gg.)* (Moscow, 1935) has data for a survey of 306 research institutes and 39 branches; it gives (pp. 158–9) a figure of 100.6 million rubles for non-budgetary funds – 48 per cent of the total expenditure figure of 209 million.
s   Figure very probably refers only to the non-budgetary funds of the research establishments under the commissariats of education.
t   Probably includes budgetary expenditure under II (2).
u   Plotnikov, *Byudzhet Sotsialisticheskogo Gosudarstva*, p. 334, states that in the First Five Year Plan 95 million rubles were spent on capital expenditure on the research institutes of the economic departments which was not included under 'science', and that in the second plan it had risen to 303 million.

SOURCES

[1]   Addition of individual items of expenditure in *Edinyi (Orientirovochnyi) Gosudarstvennyi Byudzhet Soyuza Sovetskikh Sotsialisticheskikh Respublik na 1924–1925 Byudzhetnyi God*, Part III (Moscow, 1925) pp. 71–86.
[2]   Additions of individual items of expenditure, ibid. Part IV, pp. 33–87.
[3]   *Nauchnye Dostizheniya v Promyshlennosti i Raboty Nauchno-Tekhnicheskogo Otdela VSNKh SSSR* (Moscow, 1925) p. 35.

4. Addition of individual items of expenditure from *Edinyi Gosudarstvennyi Byudzhet Soyuza Sovetskikh Sotsialisticheskikh Respublik na 1925–1926 Byudzhetnyi God. Proekt* (Moscow, 1926) pp. 83–127.
5. Addition of the figures for actual expenditure in the RSFSR, Ukraine SSR and ZSFSR budgets, for planned expenditure in the Belorussian SSR, Uzbek SSR and Turkmen SSR, *Prilozhenie k Proektu Edinogo Gosudarstvennogo Byudzheta Soyuza Sovetskikh Sotsialisticheskikh Respublik na 1925–1926 Byudzhetnyi God* (Moscow, 1926) pp. 30–554 and *Edinyi (Orientirovochnyi) ... na 1924–1925*, pp. 132–8, 269–92.
6. *Ob"yasnitel'naya Zapiska ... na 1927–1928*, p. 233.
7. Additions of individual items of expenditure from *Edinyi Gosudarstvennyi Byudzhet Soyuza Sovetskikh Sotsialisticheskikh Respublik na 1926–1927 Byudzhetnyi God. Proekt* (Moscow, 1927) pp. 63–107.
8. Addition of individual items of expenditure, ibid., pp. 221–9.
9. Addition of figures from ibid., pp. 122–3, *Ob"yasnitel'naya Zapiska .. na 1927–1928*, p. 233 and Gobza, *Vestnik Finansov*, no. 1 (1926) pp. 106–9.
10. Addition of individual items of expenditure from *Edinyi Gosudarstvennyi Byudzhet Soyuza Sovetskikh Sotsialisticheskikh Respublik na 1926–1927 Byudzhetnyi God* (Moscow, 1927) pp. 60–96.
11. Addition of individual items, ibid., pp. 152–203.
12. Addition of individual items, ibid., pp. 221–9.
13. Addition of figures from *Ob"yasnitel'naya Zapiska ... na 1927–1928*, p. 233 and *Edinyi ... na 1926–1927. Proekt*, pp. 122–3.
14. Addition of individual items from *Edinyi Gosudarstvennyi Byudzhet Soyuza Sovetskikh Sotsialisticheskikh Respublik na 1928–1929 Byudzhetnyi God. Proekt* (Moscow, 1928) pp. 54–79.
15. Addition of individual items from *Edinyi Gosudarstvennyi Byudzhet Soyuza Sovetskikh Sotsialisticheskikh Respublik na 1928–1929 Byudzhetnyi God* (Moscow, 1929) pp. 205–52.
16. *Nauchnye Kadry i Nauchno-Issledovatel'skie Uchrezhdeniya* (Moscow, 1930) p. 88; data from a survey of 1082 out of 1227 scientific establishments; this source has a figure of 62.7 million rubles for all-union and republican budgetary expenditure.
17. *Kontrol'nye Tsifry Narodnogo Khozyaistva SSSR na 1929/30* (Moscow, 1930) p. 495.
18. I. Chuev, 'Promyshlennost' v 1928/29 Godu', *Vestnik Finansov*, no. 6 (1930) p. 116.
19. *Nauchnye Kadry ...*, p. 85; the figures are estimates based on the survey figures for total expenditure of 108.5 million rubles in 1927/28, 123.6 in 1928/29.
20. K. N. Plotnikov, *Ocherki Istorii Byudzheta Sovetskogo Gosudarstva* (Moscow, 1955) p. 151.
21. Addition of individual items of expenditure in *Edinyi ... na 1929–1930*, pp. 197–212.
22. *Otchet ... za 1929–1930*, pp. 84–9.
23. V. Milyutin, 'Vtoraya Pyatiletka i Zadachi Nauchnogo Fronta', *FNIT*, no. 4–5 (1932) p. 23; this source gives the following figures for state budget expenditure: 200 million rubles in 1930, 230 in 1931 and 269 in 1932. Milyutin does not state that the 1932 figure is for planned expenditure, but it can also be found in the TsIK USSR resolution on the 1932 plan which is reprinted in *Byulleten' Finansovo-Khozyaistvennogo Zakonodatel'stva*, no. 1 (1 January 1932) p. 7.
24. *Kul'turnoe Stroitel'stvo SSSR. 1935*, pp. 258–61; data from a survey of 1100 (out of 1908) research establishments.
25. *Raskhody na Sotsial'no-Kul'turnye Meriopriyatiya po Edinomu Gosudarstvennomu Byudzhetu SSSR za 1-yu i 2-yu Pyatiletki* (Moscow, 1939) p. 21.
26. *Kul'turnoe Stroitel'stvo SSSR. 1935*, pp. 240–1.
27. Ibid., pp. 244–9.

Appendix 1 The Data for Soviet Expenditure on R & D 157

28 Ibid., pp. 258-61; figures for a survey of 1033 research establishments (out of 1582), the source gives a figure for total budgetary expenditure of 399.6 million rubles – 232.9 all-union and 166.7 republican and local.
29 *Otchet... za 1935*, pp. 5-7.
30 *Kul'turnoe Stroitel'stvo SSSR. 1935*, p. 238; this source gives figures of 322 and 210 million rubles for all-union and republican budgetary expenditure respectively.
31 *Otchet... za 1935*, pp. 58-65, 108-9.
32 Ibid., p. 212.
33 *Proekt Vtorogo Pyatiletnego Plana Razvitiya Narodnogo Khozyaistva SSSR, 1933-1937gg.*, vol. 1 (Moscow, 1934) p. 395.
34 *Otchet Narodnogo Komissariata Finansov Soyuza SSR ob Ispolnenii Edinogo Gosudarstvennogo Byudzheta Soyuza Sovetskikh Sotsialisticheskikh Respublik za 1934g.* (Moscow, 1935) p. 223.
35 *Otchet... za 1937*, p. XXV.
36 *Narodno-Khozyaistvennyi Plan Soyuza SSR na 1937 God* (Moscow, 1937) p. 35.
37 K. Ya Bauman, 'Polozhenie i Zadachi Sovetskoi Nauki', *Vestnik Akademii Nauk SSSR*, no. 10 (1936) p. 24.
38 *Kul'turnoe Stroitel'stvo SSSR* (Moscow, 1940) p. 42.
39 Ibid., p. 33.
40 A. G. Zverev, *Gosudarstvennye Byudzhety Soyuza SSR 1938-1945gg* (Moscow, 1946) p. 23.
41 Given by S. V. Kaftanov (chairman of the Committee on Higher Education) in his remarks on the proposed 1938 budget, *Vtoraya Sessiya Verkhovnogo Soveta SSSR. 10-21 Avgusta 1938g. Stenograficheskii Otchet* (Moscow, 1938) p. 129.
42 Zverev, *Gosudarstvennye Byudzhety*, p. 57.
43 *Raskhody na Sotsial'no-Kul'turnye Meriopriyatiya po Gosudarsvennomu Byudzhetu SSSR, Statisticheskii Sbornik* (Moscow, 1958) p. 42.
44 *Narodnoe Khozyaistvo SSSR v 1967g.* (Moscow, 1968) p. 888.
45 A. G. Zverev, *O Gosudarstvennom Byudzhete na 1941 God i Ispolnenii Gosudarstvennogo Byudzheta SSSR na 1939 God* (Moscow, 1941) p. 26.

TABLE A1.2 Expenditure on the industrial research establishments in the USSR 1923/24–1940[a] (million rubles, current prices)

	1923/24	1924/25	1925/26	1926/27	1927/28	1928/29	1929/30	1931[1]
I RESEARCH ESTABLISHMENTS UNDER VSNKh/NKTP SSSR								
(1) Budgetary funds								
(i) Expenditure through the category 'science'								
(a) All-Union budget	5.8[b1]	6.1[b1]	10.2[b1]	17.3[b1]	12.1[7]	13.4[d10]	22.7[24]	47.9[16]
(b) Republican budgets	—	0.3[c3]	0.5[cd4]	0.9[af6]	0.3[c8]	0.7[c11]	1.3[d25]	—
Total	—	—	—	—	—	—	—	57.4[14]
(ii) Expenditure through other budgetary categories	—	—	—	—	8.3[2]	13.2[2]	31.6[26]	72.0[14]
Total budgetary expenditure	—	—	—	—	—[g]	26.7[12]	41.0[14]	[129.4]
(2) Non-budgetary funds	0.6[2]	1.4[2]	3.1[5]	5.2[5]	14.8[9]	26.7[12]	41.0[14]	80.8[14]
(3) Total expenditure	—	—	—[e]	—	35.5[9]	58.0[13]	88.5[14]	210.2[14]

	1932	1933	1934	1935	1936	1937	1938	1939	1940
I RESEARCH ESTABLISHMENTS UNDER VSNKh/NKTP SSSR									
(1) Budgetary funds									
(i) Expenditure through the category 'science'									
(a) All-Union budget	42.6[27]	44.3[28]	39.0[15]	45.1[22]	—	57.4[32]	—	—	—
(b) Republican budgets	1.7[27]	2.1[29]	—	0.0[22]	—	—	—	—	—
Total	[44.3]	[46.4]	—	45.1[22]	—	—	—	—	—
(ii) Expenditure through other budgetary categories	—	—	37.9[20]	15.3[22]	—	—	—	—	—
Total budgetary expenditure	—	49.4[23]	230.2[23]	[60.4]	—	—	—	—	—
(2) Non-budgetary funds	189.8[h17]	172.6[23]	—	—	—	—	250[33]	—	—
(3) Total expenditure	230.8[h17]	222.0[23]	286.6[20]	>400[30]	>400[31]	—	—	>300[34]	—

The figures enclosed by [ ] are additions or subtractions of other figures in this table.
See p. 160 for notes.

TABLE A1.2 continued

	1929/30	1931[1]	1932	1933	1934	1935	1936	1937	1938	1939	1940
II ALL INDUSTRIAL RESEARCH ESTABLISHMENTS											
(1) Budgetary funds											
(i) Expenditure through the category 'science'											
(a) All-Union budget	—	—	57.8[27]	58.3[18]	52.0[19]	70.2[22]	—	85.0[32]	—	—	—
(b) Republican budgets	—	—	3.5[27]	7.3[29]	—	2.5[22]	—	—	—	—	—
Total	—	—	[61.3]	[65.6]	—	[72.7]	—	—	—	—	—
(ii) Expenditure through other budgetary categories	—	—	—	—	—	[89.4]	—	—	—	—	—
Total budgetary expenditure	—	—	—	87.8[k18]	69.7[k21]	—	—	—	—	—	—
(2) Non-budgetary funds	—	177.3[k15]	—	212.1[k18]	260.1[k21]	—	—	—	—	—	—
(3) Total expenditure	109.3[jk15]	—	—	299.8[k18]	329.8[k21]	—	—	—	—	—	—

The figures enclosed by [] are additions or subtractions of other figures in this table.
See p. 160 for notes.

160 Science and Industrialisation in the USSR

NOTES

a Before 1930 all state industry was controlled by VSNKh SSSR and the republican VSNKhy; the small amount of data which have been found for the expenditure by republican VSNKhy is included in I.
b These figures are not, in fact, the same as those for budgetary expenditure on VSNKh's research establishments which are given in the budgetary materials from which the figures for total budgetary expenditure have been reached. The sources cited in Table A1.1 for these years have figures of 5.4 million rubles in 1923/24, 3.9 in 1924/25, 7.7 in 1925/26, 8.7 in 1926/27; the differences may be attributable to the absence of some capital expenditure on research establishments.
c Expenditure only in the RSFSR, Ukraine SSR and ZSFSR.
d Planned figure.
e *FNIT*, no. 10–11 (1931) p. 3, cites a figure of 12 million rubles for expenditure on the industrial research network (it also gives a figure of 58 million for 1928/29, 257 million for 1930).
f Expenditure only in the RSFSR, Ukraine SSR, ZSFSR and Uzbek SSR.
g V. D. Esakov, *Sovetskaya Nauka v Gody Pervoi Pyatiletki* (Moscow, 1971) p. 83, gives a figure of 18.3 for state budget expenditure (administrative fund of VSNKh and *promfond*) out of a total of 35.6 million rubles.
h Estimate made towards the end of the year.
j Figure for 1930.
k Survey data.
l In 1930 the budgetary year was changed to coincide with the calendar year.

SOURCES

1 Additions of data from *Ob"yasnitel'naya Zapiska... na 1927–1928*, p. 229.
2 See Table A1.1.
3 Addition of individual items of expenditure from *Prilozhenie k Proektu ... na 1925–1926*, pp. 178–546.
4 Ibid., pp. XXX–XXXI.
5 *Ob"yasnitel'naya Zapiska... na 1927–1928*, p. 233.
6 *Edinyi... na 1926–1927*, pp. 152–3.
7 *Edinyi... na 1928–1929. Proekt*, pp. 69–70.
8 *Edinyi... na 1928–1929*, p. 212.
9 *Poyasneniya, Raschety i Obosnovaniya Vnesennykh v Obshchesoyuznuyu Rospis' Kreditov po Glavam i Paragrafam Raskhodnykh Smet, Perechni Spetsial'nykh Sredtsv i Spisok Uchrezhdenii Sostoyashchikh na Obshchesoyunom Gosudarstvennom Byudzhete v 1928/29g.* (Moscow, 1928) pp. 78–9; Kuibyshev at the Sixteenth Party Congress gave a figure of 32.5 million rubles as the expenditure on industrial research in this year, V. V. Kuibyshev, *Izbrannye Proizvedeniya* (Moscow, 1958) p. 215.
10 *Edinyi... na 1928–1929*, pp. 72–3.
11 *Edinyi... na 1929–1930*, p. 98.
12 Ibid., pp. 205–6.
13 Kuibyshev, *Izbrannye Proizvedeniya*, p. 215.
14 Esakov, *Sovetskaya Nauka...*, p. 120; the figures are taken from (archive) materials presented by NIS VSNKh to Ordzhonikidze in connection with his report to the Seventeenth Party Conference.
15 *Narodnoe Khozyaistvo SSSR* (Moscow-Leningrad, 1932) p. 551; no information is given on the coverage of the industrial research establishments – for the survey as a whole the data are for 1138 research establishments in 1930 and 1249 in 1931, out of the 1932 figure of 1565, which excludes the Uzbek SSR; expenditure on all establishments is given as 259.6 million rubles in 1930, 424.9 in 1931.
16 *Otchet... za 1931*, pp. 107–8.

Appendix 1 The Data for Soviet Expenditure on R & D

17. A. Ziskind, 'Kontrol'nye Tsifry Nauchno-Issledovatel'skikh Insitutov Tyazheloi Promyshlennosti na 1933g.', *SRIN*, no. 9–10 (1932) p. 241; this source gives a figure of 41 million rubles for budgetary expenditure.
18. *Kul'turnoe Stroitel'stvo ... (1930–1934)*, pp. 158–9; data for 143 out of 194 institutes, 52 out of 84 branches.
19. *Kul'turnoe Stroitel'stvo SSSR. 1935*, pp. 240–1.
20. A. A. Armand (ed.), *Nauchno-Issledovatel'skie Instituty Tyazheloi Promyshlennosti* (Moscow-Leningrad, 1935) p. XVIII.
21. *Kul'turnoe Stroitel'stvo SSSR. 1935*, pp. 256–61; data for 141 out of 178 institutes, 46 out of 82 branches.
22. *Otchet ... za 1935*, pp. 58–63, 91, 126–7.
23. *Nauchno-Tekhnicheskoe Obsluzhivanie Promyshlennosti. Sbornik NISa i Tekhpropa NKTP k XVII S"ezdu VKP(b)* (Moscow-Leningrad, 1934) p. 11.
24. *Otchet ... za 1929–30*, p. 78.
25. *Edinyi ... na 1929–1930*, p. 57; the planned all-union expenditure was 24.9 million rubles.
26. *Otchet ... za 1929–30*, p. 84.
27. *Otchet ... za 1932*, p. 195.
28. *Otchet ... za 1933*, pp. 90–1.
29. Ibid., p. 118.
30. L. Reinberg, 'Pokonchit's Otstavaniem Nauchno-Issledovatel'skikh Institutov Promyshlennosti', *FNIT*, no. 10 (1936) p. 101.
31. Bauman, *Vestnik Akademii Nauk SSSR*, no. 10 (1936) p. 24.
32. *Otchet ... za 1937*, p. 86.
33. Editorial in *Zavodskaya Laboratoriya*, vol. VII (1938) p. 773; it is reported in an editorial in *Industriya* 24 December 1938, that the planned expenditure for 1938 was 225 million rubles.
34. Editorial in *Industriya* 6 June 1940.

TABLE A1.3 Expenditure on 'Science' in the USSR, 1923/24–1941 (summary data and estimates: current prices, million rubles)

	Budgetary Expenditure	Non-Budgetary Funds	Total Expenditure	Index of 'Real' Expenditure* (1927/28=100)
1923/24	(16)[a]	1[b]	[17]	20
1924/25	(25)[c]	1[b]	[26]	28
1925/26	/40/[d]	—[b]	—	—
1926/27	/54/[d]	—[b]	—	—
1927/28	/81/[e]	[36]	117	100
1928/29	/103/[e]	[62]	165	130
1929/30	/216/[e]	[90]	306	219
1931	/257/[f]	/238/[f]	495	306
1932	308	[342]	(650)[g]	333
1933	344[h]	370[h]	714[h]	339
1934	434	/449/[j]	882[j]	363
1935	/678/[k]	415	[1093]	367
1936	798	(400)[l]	[1200]	335
1937	853	(500)[l]	[1350]	346
1938	(950)[m]	—	—	—
1939	(1000)	—	—	—
1940	1135	—	3000	588
(1941 plan	1032	619	1651)	

* Total expenditure deflated by index from Table A2.1

SYMBOLS USED

Figures without brackets are taken from Table A1.1.
Figures enclosed by [] are additions or subtractions from other figures in this Table.
Figures enclosed by / / are derived from the data in Tables A1.1 and A1.2 with some risk of error.
Figures enclosed by () are estimates involving substantial risk of error.

NOTES

[a] It appears that figures in Table A1.1 for budgetary expenditure in 1923/24 do not include all items of budgetary funds. It has been noted in Table A1.2 (footnote [b]) that there exist two sets of figures for all-union budgetary expenditure on the research establishments of VSNKh in this year. Using the higher figure from the later source we reach a new figure of 6.2 million rubles for all-union expenditure. The information available on the 1923/24 republican budgets is only for the RSFSR; furthermore it seems that the figure for RSFSR expenditure reached by the addition of the individual items of expenditure which are specifically stated as being scientific establishments etc., underestimates the actual expenditure. This information is taken from the proposed budget for 1924/25 and there is a large difference between 1924/25 planned and actual expenditure, particularly in the field of agriculture, with categories of expenditure listed in the report on the budget which did not appear in the plan – in the case of the RSFSR planned expenditure was 0.3 million rubles, actual 5 million. In the absence of further information I have assumed that the figure for expenditure through the RSFSR budget in 1923/24 was 40 per cent greater than the total reached by adding those items of expenditure which are recorded. On the basis of the materials for 1924/25 it is further assumed that the RSFSR spent 80 per

Appendix 1 The Data for Soviet Expenditure on R & D

cent of the total funds of the republics. Thus it is estimated that republican budgetary expenditure on science in 1923/24 was about ten million rubles. Total expenditure is, therefore, estimated to be sixteen million rubles.

b It is assumed that in 1923/24 and 1924/25 non-budgetary funds were only available to VSNKh research establishments. Not enough information exists to be able to estimate the size of these funds for 1925/26 or 1926/27.

c An estimated figure for all-union expenditure is reached (as for 1923/24) by using the higher figure for all-union budgetary expenditure on VSNKh research establishments (6.1 as opposed to 3.9 – see Table A1.2, footnote ᵇ). To reach a figure for republican budgetary expenditure it is assumed that the actual expenditure in the Belorussian, Turkmen and Uzbek republics exceeded that planned by the same proportion as for the RSFSR and the Ukraine (planned 7.6, actual 15.6); thus it is estimated that republican expenditure was sixteen million rubles and so total budgetary expenditure was twenty-four million rubles.

d Figure from Table A1.1 adjusted upwards to include later higher figure for budgetary expenditure on the research establishments of VSNKh (see Table A1.2 footnote ᵇ).

e Addition of the individual figures for all-union, republican and local budgetary expenditure on science (for 1927/28, the last is based on survey data) and of figure for budgetary expenditure outside the category 'science'.

f It is assumed that the breakdown of funds was the same for all research establishments as for the survey material for this year – 52 per cent from the budget and 48 per cent non-budgetary funds (see Table A1.1, footnote ʳ). The figure for budgetary funds thus calculated is 257 million rubles, i.e. 14 million rubles more than the figure for budgetary expenditure on 'science'.

g For total expenditure in 1932, we have a plan figure of 646 million rubles and a figure of 470 million which is given in the draft of the Second Five Year Plan. It is unclear whether this latter figure includes capital expenditure – this was to be 170 million out of the planned 646 million. Our estimate of actual expenditure is 650 million rubles, i.e. at a level similar to planned expenditure. This higher figure is suggested as a reasonable estimate for the following reasons: the lower figure is, in fact, less than the total reached by adding reported budgetary expenditure in 1932 and the non-budgetary funds allocated to the research establishments of NKTP alone (see Table A1.2) and actual total budgetary expenditure was, in fact, higher than planned (see *Otchet . . . za 1932*, p. 195). With total expenditure at 650 million rubles, actual budgetary expenditure is roughly the same proportion of this total as budgetary expenditure is of total expenditure in 1931 and 1933. It is also worth noting that the data in the draft plan on the numbers of research establishments and employment in them also vary from figures for that year in other published sources.

h Figures for a survey of research establishments: they are undoubtedly underestimates but actual expenditure may not have been much higher – the figure of 344 for funds from all budgetary sources can be compared with the figure of 324 million for budgetary expenditure on 'science' (see Table A1.1).

j It is assumed that the survey figures for non-budgetary funds (413.5 million) and for total funds (813.1 million) were the same proportion of the total as the survey figure for budgetary expenditure on science. Due to lack of data budgetary expenditure through other categories than 'science' has had to be ignored, and so the figures are probably slight underestimates.

k Additions of individual figures from Table A1.1; it is assumed that the relationship between actual and planned local budgetary expenditure was the same as for that between actual and planned all-union and republican expenditure – 562.1:532; actual local expenditure is thus estimated at 82.4 million rubles.

l Estimates based on the statement in Plotnikov, *Byudzhet Sotsialisticheskogo Gosudarstva*, p. 334, that extra-budgetary funds amounted to 2040 million rubles in the Second Five Year Plan.

m Estimates based on the statement in *Narodnoe Obrazovanie v SSSR* (Moscow, 1957) p. 780, that 3084 million rubles was spent on science through the budget in 1938–40.
n This figure is first given in the 1967 statistical handbook for the USSR as 0.3 milliard (post-1961) rubles. This means that it could, in fact, have been between 2450 and 3449 million (pre-1961) rubles. The scope of the officially published data for expenditure on science had changed from earlier years; all capital expenditure was now included and it has been suggested that some funding of defence research which was previously from the defence budget now appeared in 'science' – see C. Freeman and A. Young, *The Research and Development Effort in Western Europe, North America and the Soviet Union* (Paris, 1965) p. 121, and R. Hutchings, *Soviet Science, Technology, Design. Interaction and Convergence* (London, 1976) p. 74. On the other hand expenditure (a small amount) on some organisations – museums and exhibitions – had been excluded from science. Thus we assume that the figure for 1940 includes various expenditures not included in the contemporary information and this explains the obvious discrepancy between the 1940 figure and the figures for planned expenditure on science in 1941. Even so 1940 expenditure was possibly more likely to be below rather than above 3000 million rubles.

TABLE A1.4 Expenditure on the industrial research establishments in the USSR, 1923/24–1940 (summary data and estimates: current prices, million rubles)

	Budgetary Expenditure	Non-Budgetary Funds	Total Expenditure	Index of 'Real' Expenditure* (1927/8 = 100)
1923/24	5.8[a]	0.6	[6.4]	24
1924/25	/6.4/[ab]	1.4	[7.8]	27
1925/26	/10.7/[b]	3.1	[13.8]	43
1926/27	/18.2/[b]	5.2	[23.4]	69
1927/28	/21.5/[b]	14.8	[36.3]	100
1928/29	[31]	27	58.0	147
1929/30	(70)[c]	(60)[c]	(130)[c]	300
1931	—	—	(210)[d]	419
1932	—	—	(260)[e]	429
1933	88[f]	212[f]	300[f]	458
1934	70[f]	260[f]	330[f]	437
1935	89	—	—	—
1936	—	—	—	—
1937	85	—	—	—
1938	—	—	—	—
1939	—	—	—	—
1940	—	—	—	—

* Expenditure deflated by the index from Table A2.2

SYMBOLS USED

Figures without brackets are taken from Table A1.2.
Figures enclosed by [ ] are additions or subtractions from other figures in this Table.
Figures enclosed by / / are derived from the data in Tables A1.1 and A1.2 with some risk of error.
Figures enclosed by ( ) are estimates involving substantial risk of error.

Appendix 1 The Data for Soviet Expenditure on R & D

NOTES

a As data are not available for all republican budgets these figures may slightly underestimate budgetary expenditure: for 1923/24 data for RSFSR; in 1924/25 no data for Belorussian, Turkmen and Uzbek republics, for which, in fact, in 1925/26 no expenditure was planned.

b Additions from Table A1.2.

c The figure for total expenditure was calculated in the following way: it was assumed that the ratio of the survey data to the actual was the same as that for the survey figure for total expenditure (259.6) to the actual figure (306). To estimate the proportions of budgetary and non-budgetary establishments, the breakdown of expenditure for VSNKh establishments in that year was used.

d Calculated in the same manner as 1929/30: the survey figure for expenditure on the industrial research establishments was 177.3, that for expenditure on science 424.9 and the total expenditure figure 495 million rubles.

e It is assumed that expenditure on the industrial research establishments in 1932 was the same proportion of the estimated total expenditure (650 million rubles) as in the survey data for 1933 and 1934 (see Tables A1.1 and A1.2) i.e. 40 per cent.

f Figures from a survey of research establishments, undoubtedly underestimates.

Appendix 2 A Price Index for Soviet Expenditure on Science

The period of this study when taken as a whole was a time of rising prices and, more importantly in the present context, a time of swiftly growing (money) wages and salaries.[1] A deflator is therefore needed to obtain an estimate of the 'real' growth in expenditure; and so a price index for Soviet expenditure on science has been constructed for the years 1923/24–1940. The year 1926/27 was chosen as the base year to enable the comparison of deflated expenditure with the Soviet figures for national income which are given in 1926/27 prices. The breakdown of Soviet expenditure on science in 1926/27 is estimated to have been as follows:[2]

	per cent
Manpower costs	60
The purchase of equipment and other operational and administrative costs	25
Construction expenditure	10
The purchase of capital equipment	5
	100

The price index is based on three indices:

(a) A price index for manpower costs, which has been calculated from Soviet data for the annual average pay of all manual and office workers in the national economy (Table A2.1, column (1)); both the scattered data on rates of pay in research establishments and the crude figures obtained by dividing expenditure by the numbers employed in these establishments show a similar trend for the years for which information is available.

Appendix 2 A Price Index for Soviet Expenditure on Science

TABLE A2.1 A price index for expenditure on Science, 1923/24–1940

	Manpower costs Average annual pay of manual and office workers in the national economy		Equipment costs All Soviet machinery		Construction costs			Composite price index for Soviet expenditure on science (Manpower costs, 60% equipment costs, 30% construction costs, 10%)
	(rubles)	Index (1926/27 = 100)	(1937 = 100)	Index used (1926/27 = 100)	(1913 = 100)	(1937 = 100)	Index used (1926/27 = 100)	(1926/27 = 100)
	(1)	(2)	(3)	(4)	(5)	(6)	(7)	(8)
1923/24	[375]	60.1	—	[100]	—	—	[105]	77
1924/25	450	72.1	—	[100]	—	—	[105]	84
1925/26	571	91.5	—	[100]	269	—	104.7	95
1926/27	624	100.0	—	[100]	257	—	100.0	100
1927/28	703[a]	112.7	70	100	247	43.4[a]	95.7	107
1928/29	800[b]	128.2	68	97.1	—	43.7[b]	100.7	116
1929/30	936[c]	150.0	66	94.3	—	44.0	101.4	128
1931	1127	180.6	—	[94]	—	48.3	111.3	148
1932	1427	228.7	—	[94]	—	58.1	133.9	179
1933	1566	251.0	—	[94]	—	63.2	145.6	193
1934	1858	297.8	65	92.9	—	72.1	166.1	223
1935	2269	363.6	84	120.0	—	82.2	189.4	273
1936	2764	442.9	97	138.6	—	92.6	213.4	329
1937	3038	486.9	100	142.9	—	100	230.4	358
1938	3467	555.6	100	142.9	—	107	246.5	401
1939	[3760]	602.6	102	145.7	—	120	276.5	433
1940	4054	649.7	112	160.0	—	132	304.1	468

Figures enclosed by [] are the author's estimates.
See p. 168 for notes.

168 Science and Industrialisation in the USSR

NOTES
a Figure for 1928.
b Figure for 1929.
c Figure for 1930.

SOURCES
Column (1) 1923/24: no global figure is available; the estimate of 375 is based on the growth in pay between 1923/24 and 1924/25 in the sectors for which data does exist, see *Trud v SSSR* (Moscow, 1936) pp. 16–17.
1924/2/5–1935: idem.
1936: calculated from the previous year's figure on the basis of an index given in *SSSR i Kapitalisticheskie Strany* (Moscow-Leningrad, 1939) p. 91.
1937, 1938 and 1940: Soviet data cited in Janet G. Chapman, *Real Wages in Soviet Russia since 1928* (Cambridge, Mass., 1963) p. 109.
1939: no figure is available – it is assumed that pay was at the average of the 1938 and 1940 levels.
(2) Calculated from (1).
(3) Richard Moorsteen, *Prices and Production of Machinery in the Soviet Union, 1928–1958* (Cambridge, Mass., 1962) p. 72.
(4) Based on (3).
(5) Naum Jasny, *Soviet Prices of Producers' Goods* (Stanford, 1952) p. 71.
(6) Raymond P. Powell, *A Materials Input Index of Soviet Construction: Revised and Extended* (Santa Monica, 1959) p. 85.
(7) Based on (5) and (6).
(8) Calculated from (2), (4) and (7).

(b) A price index for equipment costs based on Moorsteen's index for Soviet machinery (Table A2.1, column (3)). It has further been assumed that prices remained stable in the years 1923/24–1927/28[3] and 1929/30–1933. This index is used as a deflator both for the expenditure on capital equipment and current expenditure on equipment, materials etc.

(c) A price index for construction costs which uses figures given by Jasny in his *Soviet Prices of Producer Goods* for 1925/26–1926/27 and Powell's price index for construction inputs for the years 1927/28–1940. Costs are assumed to have been stable in the years before 1925/26 (Table A2.1, columns (5) and (6)).

Appendix 3 The Data on Soviet R&D Manpower

The systematic collection of data on employment in science in the Soviet Union stemmed from a survey undertaken by the economic statistics sector of Gosplan in 1929[1]. Prior to this survey the only series of manpower data is for the number of scientists (*nauchnye rabotniki*) who were members of Rabpros, the trade union for education personnel[2]. The approach of the Gosplan statisticians was to add up the number of scientific posts in higher educational establishments, research establishments, the state apparatus and industry[3]. However, this survey which established the basis for the subsequent collection of data on employment in science did not include independent design and development organisations within its category of research establishments. Consequently while there is considerable published data on the manpower of research institutes and other research establishments, information on the design and development organisations which were of growing importance in industrial R&D in the 1930s is almost completely lacking.

Furthermore a problem with the data collected on the basis of a count of posts which involved the performance of scientific work was that, at this time, many people worked in more than one establishment. It was very common for a scientist to be holding both a lecturing post in higher education and a post in a research establishment; thus the figure reached by adding the number of posts, or in the survey's words, post-holders (*funktsionery*) greatly overestimated the numbers of actual scientists. It was, therefore, necessary to adjust the data for 'multiple-post-holding' (*sovmestitel'stvo*) to obtain estimates of the actual number of people.[4] The 1929 survey showed that in the USSR at that time there were 39,664 posts for scientists, of which 16,483 were in research establishments, 20,881 in higher educational establishments and 2360 in the state apparatus, industry and elsewhere.[5] However, about half of the posts in research establishments and forty per cent of those in the education sector were held by scientists who also held posts in other bodies and it is estimated that the actual number of scientists in the USSR was 30,448.[6] Staff in industrial research establishments were less likely to be holding more than one post.[7] In a 1931 survey conducted on the same basis as this 1929 study a total of 31,518 posts was converted to an estimated 22,000 scientists.[8] 'Multiple-post-holding' declined rapidly in the first half of the 1930s. In 1935 7350 pesons out of a total employment in research establishments of 153,480 are stated to be 'multiple-post-holders'.[9] Although precise information is lacking, the problem of double counting in the published data was likely to have had a minimal effect on the figures for the second half of the 1930s.

In addition to the data in contemporary statistical handbooks and materials, the handbooks published in the USSR after 1956 contain retrospective data on average annual employment in the benchmark years of 1928, 1932, 1937 and 1940. In these publications figures are given for the category 'science and scientific services' which is broken down into (i) scientific, scientific research and project-design organisations, other establishments, (ii) geological surveying organisations, and (iii) organisations of the hydro-meteorological service. It is the bodies listed in (i) which can be considered R&D organisations under the Frascati definitions. This category appears to cover not only the research establishments for which data were given in the handbooks of the time, but also the independent design and development organisations on which only scanty information is available from the earlier period – it appears that the 'other establishments of scientific services' were few in number.[10] We have no information on how the totals for the pre-war benchmark years were reached, but a particular observation that can be made with respect to the figure for 1928 is that as we have just noted the first comprehensive survey of employment in research establishments was not undertaken until 1929. Further, that survey has a figure of 40,200 for the number of 'post-holders' in research establishments[11] and project and design organisations were few at this time; consequently the modern figure of 64,000 for employment in the organisations in category (i) in 1928[12] would certainly be higher than a total estimated on the basis of data published at the time.

Notes

The citation of industrial decrees
Except where otherwise stated decrees of VSNKh SSSR are from the relevant issue of VSNKh's official journals, *Sbornik Prikazov i Tsirkulyarov po Vysshemu Sovetu Narodnogo Khozyaistva SSSR i RSFSR* (until June 1925) and *Sbornik Postanovlenii i Prikazov po Promyshlennosti* (June 1925–January 1932). Decrees of NKTP SSSR in 1932 are similarly from its *Sbornik Postanovlenii i Prikazov*. In subsequent years the periodical publication of its decrees appears to have been discontinued and where no other source appears it means that the individual decree was itself consulted.

CHAPTER 1

1. See Alexander Vucinich, *Science in Russian Culture 1861–1917* (Stanford, 1970) especially pp. 183–233.
2. Ibid., p. 303.
3. See, for example, Xenia Joukoff Eudin, Helen Dwight and Harold H. Fisher (eds), *The Life of a Chemist: Memoirs of Vladimir N. Ipatieff* (Stanford, 1946) pp. 37–8, A. S. Fedorov, *Tvortsy Nauki o Metalle* (Moscow, 1969) p. 117.
4. See, for example, the writings of the earth scientist V. I. Vernadskii, *Ocherki i Rechi Akad. V. I. Vernadskogo*, Vol. I (Petrograd, 1922) pp. 45, 151, also M. S. Bastrakova, 'Organizatsionnye Tendentsii Russkoi Nauki v Nachale XXv.', in *Organizatsiya Nauchnoi Deyatel'nosti* (Moscow, 1968) pp. 167–70.
5. These and subsequent figures are additions from data in *Otchet Gosudarstvennogo Kontrolya po Ispolneniyu Gosudarstvennoi Rospisi i Finansovykh Smet za 1913 God* (Petrograd, 1914).
6. Dumskaya Fraktsiya Progressistov, *Zamechaniya na Smetu Ministerstva Narodnogo Prosveshcheniya (po Povodu Doklada Byudzhetnoi Komissii Gosudarstvennoi Dumy)* (St Petersburg, 1914) pp. 18–22.
7. See L. Ya. Eventov, *Inostrannye Kapitaly v Russkoi Promyshlennosti* (Moscow–Leningrad, 1931), J. P. McKay, *Pioneers for Profit. Foreign Entrepreneurship and Russian Industrialisation* (Chicago, 1970).
8. McKay, *Pioneers for Profit...*, pp. 235–41.
9. Bastrakova in *Organizatsiya Nauchnoi Deyatel'nosti*, pp. 159, 176–7.
10. Vucinich, *Science in Russian Culture*, p. 381, V. V. Danilevskii, *Russkaya Tekhnika* (Leningrad, 1948) pp. 356, 359–62, 452.

11. *Ocherki i Rechi Akad. V. I. Vernadskogo*, Vol. II, p. 46.
12. Vucinich, *Science in Russian Culture*, p. 201.
13. Bastrakova in *Organizatsiya Nauchnoi Deyatel'nosti*, p. 160, Vucinich, *Science in Russian Culture*, pp. 204–13, 222–33.
14. See, for example, Vucinich, *Science in Russian Culture*, p. 228.
15. Ibid., pp. 210–13, Bastrakova in *Organizatsiya Nauchnoi Deyatel'nosti*, pp. 160–4.
16. See, for example, the experiences of the chemist Vladimir Ipatieff, Eudin, Fisher and Fisher (eds), *The Life of a Chemist*, pp. 41–67, 148.
17. Bastrakova in *Organizatsiya Nauchnoi Deyatel'nosti*, pp. 177–8, A. V. Kol'tsov, *Lenin i Stanovlenie Akademii Nauk Kak Tsentra Sovetskoi Nauki* (Leningrad, 1969) pp. 83–4, Vucinich, *Science in Russian Culture*, pp. 220–2.
18. V. I. Vernadskii, *O Gosudarstvennoi Seti Issledovatel'skikh Institutov v Rossii* (Petrograd, 1917) p. 4.
19. Cited from an archive source by Bastrakova in *Organizatsii Nauchnoi Deyatel'nosti*, p. 175.
20. *Nauchnye Kadry i Nauchno-Issledovatel'skie Uchrezhdeniya SSSR* (Moscow, 1930) p. 17; modern Soviet statistical handbooks give the total number of scientists in 1913 as 10,200 of whom 4000 were in research establishments and 6200 in higher education.
21. See Vucinich, *Science in Russian Culture*, especially pp. 273–423.
22. Ibid., p. 488.

CHAPTER 2

1. V. I. Lenin, *Polnoe Sobranie Sochinenii* (5th edition) Vol. XXXVI (Moscow, 1963) p. 263.
2. See, for example, A. V. Kol'tsov, *Lenin i Stanovlenie Akademii Nauk Kak Tsentra Sovetskoi Nauki* (Leningrad, 1969) pp. 71–7.
3. *Kommunisticheskaya Partiya Sovetskogo Soyuza v Rezolyutsiyakh i Resheniyakh S"ezdov, Konferentsii i Plenumov TsK* Vol. I (Moscow, 1954) pp. 423–4.
4. Prikaz no. 208 of the Presidium of VChK, 17 December 1919, *Iz Istorii VChK (1917–1921). Sbornik Dokumentov* (Moscow, 1958), p. 346.
5. See, for example, Xenia Joukoff Eudin, Helen Dwight and Harold H. Fisher (eds), *The Life of a Chemist: Memoirs of Vladimir N. Ipatieff* (Stanford, 1946) p. 271, Kol'tsov, *Lenin i Stanovlenie...*, pp. 137–47.
6. See R. W. Davies, 'Some Soviet Economic Controllers – I', *Soviet Studies*, XI (1960) pp. 299–300, S. I. Mokshin, *Sem' Shagov po Zemle. Ocherki o Stanovlenii i Razvitii Sovetskoi Nauki. 1917–1924* (Moscow, 1972) pp. 175–224.
7. *Organizatsiya Nauki v Pervye Gody Sovetskoi Vlasti (1917–1925). Sbornik Dokumentov* (Leningrad, 1968) pp. 300–1.
8. Ibid., p. 8.
9. *Narodnoe Prosveshchenie*, no. 18–19–20 (1921) p. 105.
10. On 1 January 1918, prices were twenty-one times the 1913 level; on 1 January 1920, 2420 times that level – R. W. Davies, *The Development of the Soviet Budgetary System* (Cambridge, 1958) p. 31.

11. For total funding of Narkompros, see Sheila Fitzpatrick, *The Commissariat of Enlightenment* (Cambridge, 1970) p. 291.
12. Ibid., pp. 256–60.
13. See, for example, the experiences of A. F. Ioffe, A. F. Ioffe, *Moya Zhizn' i Rabota* (Moscow–Leningrad, 1933) p. 21, M. S. Sominskii, *Abram Fedorovich Ioffe (1880–1960)* (Moscow–Leningrad, 1964) pp. 214–20.
14. *Organizatsiya Nauki . . . (1917–1925)*, pp. 245–50.
15. Ibid., pp. 342–4.
16. Kol'tsov, *Lenin i Stanovlenie . . .*, p. 164.
17. For the general economic background see Alec Nove, *An Economic History of the USSR* (London, 1969).
18. See R. W. Davies, 'Aspects of Soviet Investment Policy in the 1920s', in C. H. Feinstein (ed.), *Socialism, Capitalism and Economic Growth* (Cambridge, 1967) pp. 285–305.
19. *Tretii S"ezd Sovetov Soyuza Sovetskikh Sotsialisticheskikh Respublik. Stenograficheskii Otchet* (Moscow, 1925) p. 51.
20. Ibid., p. 198.
21. See N. M. Fedorovskii, 'F. E. Dzerzhinskii i Nauka', *Nauchnyi Rabotnik* (hereafter *NR*), no. 7 (1928) p. 93, and Chapter 4.
22. *Resheniya Partii i Pravitel'stva po Khozyaistvennym Voprosam v Pyati Tomakh*, Vol. I (Moscow, 1967) pp. 605–11.
23. See, for example, his speech to the November 1928 plenum of the Central Committee of the Communist Party, reported in *Ekonomicheskaya Zhizn'*, 24 November 1928.
24. They were also at different times to be directly concerned in the administration of science, see Chapters 4, 5 and 6.
25. Leon Trotsky, *Radio, Science, Technique and Society* (London, n.d.) (translation reprinted from *Labour Review*, no. 2 [1957]) p. 3.
26. N. I. Bukharin, 'Nauka i SSSR', *NR*, no. 11 (1927) p. 13.
27. In 1929 full authority was given to the commissariats themselves as regards the organisation of research institutes, Sovnarkom resolution of 6 August 1929, *Sobranie Zakonov i Rasporyazhenii Rabochego i Krest'yanskogo Pravitel'stva SSSR* (hereafter *SZ SSSR*), 1929, article 482.
28. On central government control of research, see G. I. Fed'kin, *Pravovye Voprosy Organizatsii Nauchnoi Raboty v SSSR* (Moscow, 1958) pp. 294–305.
29. *Edinyi Gosudarstvennyi Byudzhet Soyuza Sovetskikh Sotsialisticheskikh Respublik na 1927–1928 Byudzhetnyi God* (Moscow, 1928) pp. 65–6, 71.
30. See M. L. Asterman, 'Sotsial'no-Kulturnoe Stroitel'stvo (Ocherk Perpektivnogo Postroeniya)', *Planovoe Khozyaistvo*, no. 12 (1927) p. 127.
31. In 1923/24 industrial wholesale prices were over twice 1913 levels and the average wage in industry roughly fifty per cent higher than in 1913. Eugène Zaleski, *Planning for Economic Growth in the Soviet Union 1918–1932* (Chapel Hill, 1971) pp. 390, 396, 398.
32. See Chapter 4.
33. *Pyatnadtsatyi S"ezd VKP(b): Dekabr' 1927 Goda. Stenograficheskii Otchet*, Vols. I and II (Moscow, 1961 and 1962).

34. Ibid., Vol. II pp. 1001–2.
35. Ibid., Vol. II p. 988.
36. Ibid., Vol. II p. 1136.
37. Ibid., Vol. II pp. 1451–2.
38. Ibid., Vol. II p. 1436.
39. For further details of these events see Chapter 5.
40. This is a reference to the eighteenth-century Russian landowners' construction of libraries for books which they never read; the Neskuchnyi Palace was the library built by Orlov-Chesmenskii, a famous horse-breeder.
41. *Shestnadtsataya Konferentsiya VKP (b). Aprel' 1929 Goda. Stenograficheskii Otchet* (Moscow, 1962) p. 247.
42. *Pyatiletnii Plan Narodno-Khozyaistvennogo Stroitel'stva SSSR* (2nd edition) (Moscow, 1930) Vol. II, Part I, pp. 161–2, 201–5, 311–12, 484–5, Part II, p. 19.
43. *5 S"ezd Sovetov. Stenograficheskii Otchet* (Moscow, 1929) Bulletin, 8, p. 40, Bulletin 14, p. 8.
44. For the capital expenditure figures see E. H. Carr and R. W. Davies, *Foundations of a Planned Economy 1926–1929* Vol. I (London, 1969) p. 1035 and Zaleski, *Planning for Economic Growth* ..., p. 246.
45. *Trud v SSSR* (Moscow, 1936) p. 19.
46. Nove, *An Economic History* ..., p. 224.
47. *Kul'turnoe Stroitel'stvo SSSR* (Moscow, 1940) p. 230, *Sotsialisticheskoe Stroitel'stvo SSSR* (Moscow, 1934) p. 420.
48. Its findings with regard to the industrial research network are reported in *Za Industrializatsiyu* (hereafter *Za Ind.*) 22 December 1932.
49. *Sotsialisticheskoe Stroitel'stvo* (1934), p. 420.
50. *Proekt Vtorogo Pyatiletnogo Plana Razvitiya Narodnogo Khozyaistva SSSR (1933–1937 gg.)* (Moscow, 1934), Vol. I, p. 538.
51. *Kul'turnoe Stroitel'stvo SSSR* (1940) p. 230.
52. *Forty Years of Soviet Power* (Moscow, 1958) p. 259, *Kul'turnoe Stroitel'stvo SSSR v Tsifrakh: ot VI k VII S"ezdu Sovetov (1930–1934 gg.)* (Moscow, 1935) p. 143.
53. For two examples in Dnepropetrovsk see J. G. Crowther, *Soviet Science* (London, 1936) p. 142.
54. Approved by Narkompros RSFSR on 30 March 1932, reprinted in *Byulleten' Finansovogo i Khozyaistvennogo Zakondatel'stva* no. 27–28 (13 June 1932) pp. 64–5.
55. See, for example, a series of short articles in *Front Nauki i Tekhniki* (hereafter *FNIT*), no. 12 (1933) pp. 45–58.
56. For details of the calculation based on the published breakdown see R. A. Lewis, *Industrial Research and Development in the USSR 1924–1935* (unpublished PhD thesis: University of Birmingham, 1975) pp. 363–4.
57. A recent article gives a figure of 1.5 million rubles for contract work undertaken by Leningrad University in 1932, but states that the majority was for analytical and computing work, K. Ya.Kondrat'ev and L. A. Shilov, 'Chto Daet Khozyaistvennyi Dogovor Vuzu i Proizvodstvu', *Vestnik Vysshei Shkoly*, no. 5 (1967) p. 48.

58. See below, Chapter 5.
59. Sovnarkom decree of 11 November 1937, reprinted in *Finansovoe i Khozyaistvennoe Zakondatel'stvo*, no. 34–35 (20 December 1937) pp. 19–21.
60. See, for example, the article by S. V. Kaftanov, chairman of the Committee for Higher Education which was attached to Sovnarkom partly for the purpose of stimulating research, 'Itogi Godovoi Raboty i Zadachi Vysshei Shkoly', *Sovetskaya Nauka*, no. 7 (1939) p. 127.
61. *Narodnoe Khozyaistvo SSSR. Statisticheskii Sbornik* (Moscow, 1956) p. 233.
62. *Proekt Vtorogo Pyatiletnego Plana* ..., Vol. I, p. 395.
63. See, for example, the report of Kuibyshev to the Seventeenth Party Congress at the beginning of 1934, *XVII S"ezd Vsesoyuznoi Kommunisticheskoi Partii (b). Stenograficheskii Otchet* (Moscow, 1934) p. 399.
64. See Appendix 1, Table A1.3, footnote ^{n}.
65. *Bol'shaya Sovetskaya Entsiklopediya* (2nd edition) Vol. XXIX, p. 302.
66. See Appendix 1 and the appendix by R. W. Davies, G. R. Barker and R. Fakiolas, in C. Freeman and A. Young, *The Research and Development Effort in Western Europe, North America and the Soviet Union* (Paris, 1965) pp. 116–23.
67. *Otchet ob Ispolnenii Gosudarstvennogo Byudzheta SSSR za 1935 God* (Moscow, 1937); for further details of the calculation see Lewis, *Industrial Research* ..., pp. 361–8.
68. *Sotsialisticheskoe Stroitel'stvo Soyuza SSR (1933–1938 gg.)* (Moscow–Leningrad, 1939) p. 18.
69. See Appendix 2.
70. Further information on these bodies is given in Chapter 3.
71. Naum Jasny, *The Soviet Economy during the Plan Era* (Stanford, 1951) p. 22.
72. Vannevar Bush, *Science: The Endless Frontier* (Washington, 1945) p. 80; Soviet National Income was, of course, much smaller — perhaps three to four times — than that of the United States.
73. Ibid.
74. *Sotsialisticheskaya Rekonstruktsiya i Nauchno-Issledovatel'skaya Rabota. Sbornik Nauchno-Issledovatel'skogo Sektora PTEU VSNKh SSSR k XVI S"ezdu VKP(b)* (Moscow, 1930) p. 4.
75. I. S. Samokhvalov, 'Chislennost' i Sostav Nauchnykh Rabotnikov SSSR', *Sotsialisticheskaya Rekonstruktsiya i Nauka* (hereafter *SRIN*), no. 1 (1934) p. 134.
76. *Kul'turnoe Stroitel'stvo SSSR. 1935* (Moscow, 1936) pp. 230–1.
77. Out of the 300 scientists in the All-Union Electrical Engineering Institute in 1934, 230 had completed higher education after 1928, *Nauchno-Tekhnicheskoe Obsluzhivanie Tyazheloi Promyshlennosti. Sbornik NISa i Tekhpropa NKTP k XVII S"ezdu VKP(b)* (Moscow–Leningrad, 1934) p. 95.
78. See, for example, comments on the auxiliary personnel at the Ukrainian Physical Technical Institute in Lucie Street (ed.), *I Married a Russian; Letters from Khar'kov* (London, 1944) pp. 95, 98, 109.

176 *Science and Industrialisation in the USSR*

79. See, for example, E. A. Chudakov, 'Problemy Nauchno-Issledovatel'skoi Raboty v Oblasti Mashinostroeniya', *Sovetskaya Nauka*, no. 4 (1939) p. 67.
80. G. Kharat'yan, 'Chislennost' i Sostav Nauchnykh i Nauchno-Pedagogicheskikh Kadrov v SSSR', *Vestnik Statistiki*, no. 4 (1962) p. 64.
81. For remarks on well-endowed institutes, see Crowther, *Soviet Science*, pp. 213–14, 226, 241, 257.
82. See, for example, the speech of Pokrovskii to the Fifth Congress of Soviets, *5 S″ezd Sovetov*..., Bulletin 11, p. 30.
83. *Pyatiletnii Plan Nauchno-Eksperimental'noi Raboty v Svyazi s Rekonstruktsiei Promyshlennosti SSSR*, no. 8, *Pyatiletnii Plan Rabot Instituta Chistykh Reaktivov* (Moscow, 1929) p. 4.
84. P. P. Budnikov, 'Nauchno-Issledovatel'skaya Rabota v Khimicheskikh Vuzov', *Sovetskaya Nauka*, no. 5 (1940) pp. 144–5; V. V. Longinov, 'Eshche Raz o Reaktivakh', *Zavodskaya Laboratoriya* (hereafter *ZL*), no. 5 (1937) pp. 535–40.
85. See Crowther, *Soviet Science*, pp. 81–3, M. Ruhemann, *New Scientist* 2 November 1967, pp. 276–7.
86. For further details see Chapter 3.
87. For the case of aeroplane design see Chapter 9 and A. Sharagin (G. A. Ozerov), *Tupolevskaya Sharaga* (Frankfurt/M, 1971).

CHAPTER 3

1. For further details see Chapter 4.
2. See, for example, the speech by Kuibyshev to the First All-Union Conference for the Planning of Scientific Research in 1931, V. V. Kuibyshev, *Nauke–Sotsialisticheskii Plan* (Moscow–Leningrad, 1931) p. 14.
3. *Nauchnye Dostizheniya v Promyshlennosti i Raboty Nauchno-Tekhnicheskogo Otdela VSNKh SSSR* (Moscow, 1925) p. 34.
4. Decree of Sovnarkom SSSR of 7 August 1928, reprinted in *Resheniya Partii i Pravitel'stva po Khozyaistvennym Voprosam v Pyati Tomakh*, Vol. I (Moscow, 1967) pp. 750–5.
5. For these and subsequent data on the numbers of industrial research establishments, see A. Ziskind, 'Naucho-Issledovatel'skie Kadry v Promyshlennosti', *FNIT*, no. 7–8 (1932) p. 105, *Sotsialisticheskoe Stroitel'stvo SSSR* (Moscow, 1934) p. 420, *Kul'turnoe Stroitel'stvo SSSR* (Moscow, 1940) p. 238.
6. L. Reinberg, 'Pokonchit' s Otstavaniem Nauchno-Issledovatel'skikh Institutov Promyshlennosti', *FNIT*, no. 10 (1936) pp. 100–6.
7. *Torgovo-Promyshlennaya Gazeta* (hereafter *TPG*) 5 October 1926.
8. N. Mezenin, *Metallurg Grum-Grzhimailo* (Moscow, 1977) pp. 95–8; *TPG* 27 November 1926.
9. *Predpriyatiya, Khozorgany i Uchrezhdeniya Narodnogo Komissariata Tyazheloi Promyshlennosti* (Moscow-Leningrad, 1935); in this handbook of the organisations under NKTP, the commissariat for heavy

industry, some branches are excluded, presumably for reasons of secrecy; there is, for example, no information on factories and other organisations in the aircraft industry. Information on the number of design organisations in the aircraft industry in the mid-1930s is given in *Istoriya Vtoroi Mirovoi Voiny 1939-1945*, Vol. II (Moscow, 1974) p. 195.

10. See Chapter 9.
11. *Predpriyatiya, Khozorgany...*
12. See, for example, I. Polyakov, *Industriya*, 10 December 1938.
13. S. S. Khromov, *F. E. Dzerzhinskii vo Glave Metallopromyshlennosti* (Moscow, 1966) p. 232; Gipro – an acronym for State Institute for Projecting – was to be widely used in the title of such organisations.
14. V. S. Lel'chuk, *Sozdanie Khimicheskoi Promyshlennosti SSSR. Iz Istorii Sotsialisticheskoi Industrializatsii* (Moscow, 1964) p. 86; it was later to be renamed Giprokhim.
15. *Prepriyatiya, Khozorgany...*
16. S. Mosin, *Industriya* 21 November 1938.
17. Alcan Hirsch, *Industrialised Russia* (New York, 1934) pp. 30-1.
18. S. Mosin, *Industriya* 21 November 1938.
19. See Sovnarkom decree of 11 November 1937 on posts and pay in higher education, reprinted in *Finansovoe i Khozyaistvennoe Zakondatel'stvo*, no. 34-35 (20 December 1937) pp. 19-21; and Chapter 5.
20. *Industriya* 24 December 1938.
21. Zh. S. Beilin, 'Nauchno-Issledovatel'skaya Rabota v Vtuzakh Mashinostroeniya', *Sovetskaya Nauka*, no. 6 (1939) p. 123.
22. For the example of the higher educational establishments of the chemical industry see S. Ya. Plotkin, 'Vysshee Khimiko-Tekhnologicheskoe Obrazovanie v SSSR', *Sovetskaya Nauka*, no. 3-4 (1940) p. 154.
23. See, for example, E. A. Chudakov, 'Problemy Nauchno-Issledovatel'skoi Raboty v Oblasti Mashinostroeniya', *Sovetskaya Nauka*, no. 4 (1939) p. 70.
24. For greater detail see Chapter 8.
25. *Industriya* 24 December 1938.
26. *Istoriya Moskovskogo Avtozavoda imeni I. A. Likacheva* (Moscow, 1966) pp. 214-16.
27. S. Kostyuchenko, I. Khrenov and Yu. Fedorov, *Istoriya Kirovskogo Zavoda 1917-1945* (Moscow, 1966) pp. 424-9, 545-7, 556-60.
28. Julian M. Cooper, *The Development of the Soviet Machine Tool Industry* (unpublished PhD thesis: University of Birmingham, 1975) p. 308.
29. Chudakov, *Sovetskaya Nauka*, no. 4 (1939) p. 71.
30. See Chapter 9.
31. See A. N. Krylov, *Moi Vospominaniya* (Moscow, 1945) pp. 347-52.
32. For an account of the relations between the Academy and the government in these years see Loren R. Graham, *The Soviet Academy of Sciences and the Communist Party 1927-1932* (Princeton, 1967).
33. N. M. Mitryakova, 'Struktura, Nauchnye Uchrezhdeniya i Kadry AN SSSR', in *Organizatsiya Nauchnoi Deyatel'nosti* (Moscow, 1968) p. 214.
34. G. A. Knyazev and A. V. Kol'tsov, *Kratkii Ocherk Istorii Akademii Nauk SSSR* (Moscow-Leningrad, 1964) p. 100.

178 *Science and Industrialisation in the USSR*

35. Mitryakova in *Organizatsiya Nauchnoi Deyatel'nosti*, p. 215.
36. See Chapter 6.
37. I. P. Bardin (ed.), *Ocherki po Istorii Akademii Nauk. Tekhnicheskie Nauki* (Moscow–Leningrad, 1945) p. 15.
38. For United States manpower see George Perazich and Philip M. Field, *Industrial Research and Changing Technology* (Philadelphia, 1940) pp. 56–7, 64.
39. Ibid., p. 78, and Table 3.1.
40. Institutes and branches have been counted as separate establishments; manpower data from Table 3.1, and for numbers of establishments see *Kul'turnoe Stroitel'stvo SSSR. 1935* (Moscow, 1936) pp. 222–5.
41. Perazich and Field, *Industrial Research and Changing Technology*, p. 65.
42. Ibid., p. 66.
43. In 1933 the 1562 companies which reported research activity had 1584 laboratories, ibid., p. 63.
44. E. H. Carr and R. W. Davies, *Foundations of a Planned Economy 1926–1929*, Vol. I (London, 1969) pp. 431–52.
45. *Resheniya Partii* . . . , p. 752.
46. *Prikaz VSNKh SSSR*, no. 1027, 26 July 1929.
47. Ziskind, *FNIT*, no. 7–8 (1932) p. 105.
48. See, for example, A. Ziskind, 'Organizatsiya Nauchno-Issledovatel'skoi Raboty v Promyshlennosti', *FNIT*, no. 6 (1931) p. 51.
49. Data on the distribution of scientists from *Nauchnye Kadry i Nauchno-Issledovatel'skie Uchrezhdeniya SSSR* (Moscow, 1930) p. 20, V. D. Esakov, *Sovetskaya Nauka v Gody Pervoi Pyatiletki* (Moscow, 1971) p. 119; for the distribution of the industrial labour force in 1932, see *Sotsialisticheskoe Stroitel'stvo* (1934) pp. 310–11.
50. *Direktivy i Formy po Sostavleniyu Tematicheskikh Planov i Kontrol'nyikh Tsifr na 1932g. Nauchno-Issledovatel'skikh Uchrezhdenii VSNKh* (Moscow-Leningrad, 1931) p. 3.
51. *XVII Konferentsiya Vsesoyuznoi Kommunisticheskoi Partii (b). Stenograficheskii Otchet* (Moscow, 1932) p. 78.
52. *Sotsialisticheskoe Stroitel'stvo* (1934) pp. 419–20.
53. *Kul'turnoe Stroitel'stvo SSSR. 1935*, pp. 222–5, *Kul'turnoe Stroitel'stvo SSSR* (Moscow, 1940) p. 233.
54. *Nauchnye Kadry* . . . , p. 20; *Kul'turnoe Stroitel'tsvo SSSR* (1940) p. 233.
55. *Predpriyatiya, Khozorgany* . . .
56. See Chapter 8.
57. *'ENIMS' i Ordena Lenina Zavoda 'Stankokonstruktsiya'. 25 let* (Moscow, 1958), A. A. Armand (ed.), *Nauchno-Issledovatel'skie Instituty Tyazeloi Promyshlennosti* (Moscow-Leningrad, 1935) pp. 591–8.
58. Ibid., p. 520.
59. Crowther, *Soviet Science*, p. 219, V. G. Prelkov, 'Voprosy Elektrifikatsii Promyshlennosti i Transporta: iz Rabot Vsesoyuznogo Elektrotekhnicheskogo Instituta 1932g.', *SRIN*, no. 4 (1933) pp. 150–61.
60. Crowther, *Soviet Science*, p. 257.
61. Armand (ed.), *Nauchno-Issledovatel'skie Instituty* . . . , p. 45.

62. Ibid., pp. 203, 211.
63. Crowther, *Soviet Science*, p. 219.
64. John Erickson, *The Soviet High Command* (London, 1962) pp. 252–79, 331–49.
65. Lel'chuk, *Sozdanie Khimicheskoi Promyshlennosti* ..., pp. 115–16; however, the author of a recent western study would undoubtedly strongly dispute this statement – Anthony C. Sutton, *Western Technology and Soviet Economic Development 1917 to 1930* (Stanford, 1968) pp. 209–24 and *Western Technology and Soviet Economic Development 1930–1945* (Stanford, 1971) pp. 97–114.
66. D. A. Gerasimov, 'Nauka i Ee Primenenie v Torfyanoi Promyshlennosti', *SRIN*, no. 2 (1934) p. 79.
67. For an account of its development see R. A. Lewis, 'Innovation in the USSR: The Case of Synthetic Rubber', *Slavic Review* (forthcoming).
68. See P. M. Luk'yanov, *Kratkaya Istoriya Khimicheskoi Promyshlennosti SSSR: ot Voznikoveniya Khimicheskoi Promyshlennosti v Rossii do Nashikh Dnei* (Moscow, 1959) pp. 326–9.
69. See Chapter 9.
70. Eleven in 1935, see Armand (ed.), *Nauchno-Issledovatel'skie Instituty* ...; basic research was also being done in the research establishments of other bodies, such as the Academy of Sciences and Narkompros.
71. Crowther, *Soviet Science*, pp. 192–3.
72. J. G. Crowther, *Science in Soviet Russia* (London, 1931) p. 61; for another example, see Armand (ed.), *Nauchno-Issledovatel'skie Instituty* ..., pp. 577–82.
73. See Chapter 9.
74. Crowther, *Soviet Science*, pp. 100–1, Armand (ed.), *Nauchno-Issledovatel'skie Instituty* ..., p. 28.
75. *ZL*, no. 5 (1937) p. 534.
76. *Pyatiletnii Plan Nauchno-Eksperimental'noi Raboty v Svyazi s Rekonstruktsiei Promyshlennosti SSSR*, no. 8, *Pyatiletnii Plan Rabot Instituta Chistykh Reaktivov* (Moscow, 1929) p. 4, Armand (ed.), *Nauchno-Issledovatel'skie Instituty* ..., p. 160.
77. Armand (ed.), *Nauchno-Issledovatel'skie Instituty* ..., pp. 716–27.
78. Crowther, *Soviet Science*, p. 252.
79. Ziskind, *FNIT*, no. 7–8 (1932) p. 106, *Kul'turnoe Stroitel'stvo SSSR* (1940) p. 242.
80. *The Measurement of Scientific and Technical Activities: Proposed Standard Practice for Surveys of Research and Experimental Development* (Paris, 1970).
81. *Industrial Research* (New York, 1969).
82. Ibid., pp. 18, 20.
83. See, for example, Armand (ed.), *Nauchno-Issledovatel'skie Instituty* ..., and *Nauchno-Tekhnicheskoe Obsluzhivanie Tyazheloi Promyshlennosti. Sbornik NISa i Tekhpropa k XVII S"ezdu VKP(b)* (Moscow-Leningrad, 1934).

CHAPTER 4

1. *Sobranie Ukazonenii i Rasporyazhenii Rabochego i Krest'yanskogo Pravitel'stva RSFSR* (hereafter *SU RSFSR*), 1917–18, article 183, 24 December 1917.
2. *Organizatsiya Nauki v Pervye Gody Sovetskoi Vlasti (1917–1925). Sbornik Dokumentov* (Leningrad, 1968) p. 21–3.
3. See N. P. Gorbunov's contribution to *V. I. Lenin vo Glave Velikogo Stroitel'stva* (Moscow, 1960) p. 178.
4. See Gorbunov's letter to Lenin which was reprinted in *Novyi Mir*, no. 8 (1964) pp. 278–9.
5. M. S. Bastrakova, *Stanovlenie Sovetskoi Sistemy Organizatsii Nauki (1917–1922)* (Moscow, 1973) pp. 99–103.
6. See, for example, *Organizatsiya Nauki ... (1917–1925)*, pp. 150–2 and Sheila Fitzpatrick, *The Commissariat of Enlightenment* (Cambridge, 1970) p. 71.
7. See A. V. Kol'tsov, *Lenin i Stanovlenie Akademii Nauk Kak Tsentra Sovetskoi Nauki* (Leningrad, 1969) pp. 69–80.
8. Fitzpatrick, *the Commissariat of Enlightenment*, p. 71.
9. Ibid., p. 72, Kol'tsov, *Lenin i Stanovlenie* ..., pp. 58–64.
10. Bastrakova, *Stanovlenie Sovetskoi Sistemy* ..., pp. 166–7.
11. *Organizatsiya Nauki ... (1917–1925)*, p. 81.
12. *SU RSFSR*, 1917–18, article 617.
13. *SU RSFSR*, 1919, article 161, 20 August 1919.
14. Bastrakova, *Stanovlenie Sovetskoi Sistemy* ..., p. 258.
15. *Organizatsiya Nauki ... (1917–1925)*, pp. 157–8, *Pyatiletnii Plan Nauchno-Eksperimental'noi Raboty v Svyazi s Rekonstruktsiei Promyshlennosti SSSR*, no. 23, *Pyatiletnii Plan Rabot Gosudarstvennogo Keramicheskogo Issledovatel'skogo Instituta* (Moscow, 1929) p. 10.
16. See *Organizatsiya Nauki ... (1917–1925)*, pp. 318–21, 329–32, 335–6.
17. Xenia Joukoff Eudin, Helen Dwight and Harold H. Fisher (eds), *The Life of a Chemist: Memoirs of Vladimir N. Ipatieff* (Stanford, 1946), p. 359, M. Ya. Lapirov-Skoblo (ed.), *Rabota Nauchno-Tekhnicheskikh Uchrezhdenii Respubliki 1918–1919* (Moscow, [1919]) p. 9.
18. *Organizatsiya Nauki ... (1917–1925)*, pp. 96, 101, *Materialy Pervogo Vsesoyuznogo S''ezda po Voprosam Khimicheskoi Promyshlennosti* (Moscow, 1930) pp. 207–8.
19. A. Lapis, 'Itogi i Blizhaishie Perspektivy Raboty UNTO', *Byulleten' Nauchno-Tekhnicheskogo Otdela pri Ukrsovnarkhoze*, no. 4–5 (1921) p. 5.
20. Lapirov-Skoblo (ed.), *Rabota ... 1918–1919*, p. 9.
21. *Organizatsiya Nauki ... (1917–1925)*, pp. 89–91.
22. See the statute of NTO VSNKh of 19 December 1923, reprinted ibid., pp. 98–100.
23. Eudin, Fisher and Fisher (eds), *The Life of a Chemist*, p. 288.
24. M. Ya. Lapirov-Skoblo (ed.), *Rabota Nauchno-Tekhnicheskikh Uchrezhdenii v 1920 Godu* (Moscow, 1920) p. II.
25. See Gorbunov's letter to Lenin which was reprinted in *Novyi Mir*, no. 8 (1964) p. 279.

26. See the December 1923 statute, reprinted in *Organizatsiya Nauki*... *(1917–1925)*, pp. 98–100.
27. For this and subsequent material on the administration of industry, see Alexander Baykov, *The Development of the Soviet Economic System* (Cambridge, 1946); E. H. Carr and R. W. Davies, *Foundations of a Planned Economy 1926–1929*, Vol. I (London, 1969); A. M. Rubin, *Organizatsiya Upravleniya Promyshlennost'yu v SSSR (1917–1967)* (Moscow, 1969); G. Sakharov, N. Chernai and O. Kabakov, *Ocherki Organizatsii Tyazheloi Promyshlennosti SSSR* (Moscow, 1934); and A. V. Venediktov, *Organizatsiya Gosudarstvennoi Promyshlennosti v SSSR*, Vol. II, *1921–1934* (Leningrad, 1961).
28. Eudin, Fisher and Fisher (eds), *The Life of a Chemist*, pp. 359, 363.
29. Ibid., pp. 361–2.
30. *Organizatsiya Nauki*... *(1917–1925)*, pp. 98–100.
31. V. Glebova, *TPG* 12 November 1924.
32. Ibid.
33. A. N. Bakh, *TPG* 19 November 1924.
34. See remarks by Bogdanov, head of VSNKh RSFSR, *TPG* 26 November 1924.
35. M. Ya. Lapirov-Skoblo, *TPG* 4 December 1924.
36. *TPG* 28 January 1925.
37. *TPG* 6 February 1925.
38. No full version has been found; the issue of *TPG* for 6 February 1925 and the theses published in *Nauchnye Dostizheniya v Promyshlennosti i Raboty Nauchno-Tekhnicheskogo Otdela VSNKh SSSR* (Moscow, 1925) pp. 34–9 provide the material used here.
39. For details of the roles of these bodies, see *Organizatsiya Nauki*... *(1917–1925)*, pp. 293–6, 333–8; no information on the actual operation of these bodies has been found.
40. *TPG* 6 February 1925.
41. Ibid.
42. Ibid.
43. His speech was reprinted in *Nauchnye Dostizheniya*..., pp. 40–4.
44. *TPG* 6 February 1925.
45. On the importance of Dzerzhinskii's views for NTO in the period until his death in 1926, see Yu. N. Flakserman, 'Pamyati F. E. Dzerzhinskogo', *Tekhniko-Ekonomicheskii Vestnik*, no. 8 (1926) p. 495.
46. Yu. N. Flakserman, 'Puti Stroitel'stva Nauchnoissledovatel'skikh Institutov', *Tekhniko-Ekonomicheskii Vestnik*, no. 8 (1926) pp. 525–7.
47. Carr and Davies, *Foundations of a Planned Economy*, Vol. I, pp. 334–8.
48. Cited in G. V. Kuibysheva, O. A. Lezhava, N. V. Nelidov and A. F. Khavin, *Valerian Vladimirovich Kuibyshev* (Moscow, 1966) p. 309.
49. *Prikaz VSNKh SSSR*, no. 1019, 4 September 1926, reprinted in *TPG* 5 September 1926.
50. Carr and Davies, *Foundations of a Planned Economy*, Vol. I, p. 355.
51. *Prikaz VSNKh SSSR*, no. 864, 27 May 1925.
52. *Prikaz VSNKh SSSR*, no. 416, 24 February 1926.
53. Eudin, Fisher and Fisher (eds), *The Life of a Chemist*, pp. 412–24; see also Leon Trotsky, *My Life* (New York, 1930) p. 518.
54. *Prikaz VSNKh SSSR*, no. 51, 15 October 1926.

55. *Prikaz VSNKh SSSR*, no. 279, 2 December 1926, reprinted in *TPG* 3 December 1926.
56. Trotsky was removed as a result of the struggle inside the Communist Party; Pyatakov, one of his supporters, lost his post of deputy commissar. Ipatieff reports that he had anticipated a change for the worse on the death of Dzerzhinskii and suggests a personal antipathy between himself and Kuibyshev which made him expect an early dismissal. Eudin, Fisher and Fisher (eds), *The Life of a Chemist*, pp. 425–6.
57. The brother of Ya. M. Sverdlov, the first titular head of the Soviet government, who had died in 1918; Ipatieff had a very low opinion of his capabilities, ibid., p. 432.
58. See Appendix 1, Table A1.2.
59. *Ob"yasnitel'naya Zapiska k Proektu Edinogo Gosudarstvennogo Byudzheta Soyuza Sovetskikh Sotsialisticheskikh Respublik na 1927–1928 Byudzhetnyi God* (Moscow, 1928) p. 233.
60. *Pyatnadtsatyi S"ezd VKP(b). Dekabr' 1927 Goda. Stenograficheskii Otchet*, Vol.II (Moscow, 1962) p. 1462.
61. V. D. Esakov, *Sovetskaya Nauka v Gody Pervoi Pyatiletki* (Moscow, 1971) pp. 81–3.
62. See, for example, Carr and Davies, *Foundations of a Planned Economy*, Vol.I, pp. 431–52, 898–915.
63. *Protokol VSNKh SSSR*, 1927–8, article 403, 19 May 1928; Kuibysheva *et al.*, *Valerian Vladimirovich Kuibyshev*, pp. 310, 317.
64. See *Protokol VSNKh SSSR*, 1927–8, article 403, 19 May 1928; R. W. Davies, 'Aspects of Soviet Investment Policy in the 1920s', in C. H. Feinstein (ed.), *Socialism, Capitalism and Economic Growth* (Cambridge, 1967) pp. 292–4.
65. Esakov, *Sovetskaya Nauka* ..., pp. 83–6.
66. Ibid., pp. 88–90.
67. Resolution of Sovnarkom of 7 August 1928, reprinted in *Resheniya Partii i Pravitel'stva po Khozyaistvennym Voprosam v Pyati Tomakh*, Vol. I (Moscow, 1967) pp. 750–5.
68. However, no reference has been found to the work of this body so it may not have become fully operational.
69. *Prikaz VSNKh SSSR*, no.969, 18 August 1928.
70. *Prikaz VSNKh SSSR*, no. 713, 28 May 1928.
71. *Protokol VSNKh SSSR*, 1928–9, article 85, 5 December 1928; the transfer was put into effect in January 1929, *Prikaz VSNKh SSSR*, no. 314, 10 January 1929.
72. Decree of Sovnarkom of 28 April, reprinted in *Resheniya Partii* ..., Vol. I, pp. 720–4.
73. See above, note 33.
74. The petition was reprinted in *Pervyi Plenum Komiteta po Khimizatsii Narodnogo Khozyaistva SSSR pri SNK SSSR. 3–6 Sentyabrya 1928* (Leningrad, 1930) pp. 335–41.
75. Decree of Sovnarkom of 12 July 1929, reprinted in *Pervyi Plenum* ..., pp. 345–8.
76. *5 S"ezd Sovetov. Stenograficheskii Otchet* (Moscow, 1929) Bulletin 11, p. 30.

Notes to pp. 46–56 183

77. A. F. Khavin, *U Rulya Industrii* (Moscow, 1968) p. 146.
78. *Kommunisticheskaya Partiya Sovetskogo Soyuza v Rezolyutsiyakh i Resheniyakh S"ezdov, Konferentsii i Plenumov TsK*, Vol. I (Moscow, 1954) pp. 597–8.
79. See an article by G. Sakharov in *TPG* 4 September 1929.
80. *TPG* 5 September 1929.
81. *TPG* 8 September 1929; the report of the discussion which follows comes from this source.
82. The decentralisation would already appear to have started; it is reported in *TPG* 14 June 1929 that the All-Union Electrical Engineering Institute had been transferred from NTU to the State Electrical Trust.
83. *Prikaz VSNKh SSSR*, no. 795, 31 May 1929.
84. He was appointed by *Prikaz VSNKh SSSR*, no. 1021, 8 September 1928.
85. N. I. Bukharin, 'Nauka i SSSR', NR, no. 11 (1927) pp. 6–16.
86. *Reorganizatsiya Upravleniya Promyshlennost'yu* (Moscow, 1929) p. 22.
87. *TPG* 12 September 1929.
88. *Ekonomicheskaya Zhizn'* 13 October 1929; on this occasion he was supported by Bakh.
89. *Reorganizatsiya Upravleniya . . .*, pp. 25–40.
90. *Resheniya Partii . . .*, Vol. II (Moscow, 1967) pp. 141–2.
91. *Bol'shevik*, no. 23–24 (1929), pp. 64, 70.
92. *Prikaz VSNKh SSSR*, no. 508, 30 December 1929; Bukharin went on a month's leave on 16 December, being temporarily replaced by Sverdlov, *Prikaz VSNKh SSSR*, no. 417, 16 December 1929.
93. One of these, the Institute for the Study of the North was dissolved as a result of the investigation, *Postanovlenie Prezidiuma VSNKh SSSR*, Protokol no. 3 (1930), 348; the final fate of the other, the North Caucasus Industrial Institute, is unknown.
94. That the number of institutes referred to in the decree is less than the total number which seem to have been in existence at the time can probably be explained by the transfer of other institutes, such as the All-Union Electrical Engineering Institute (see note 82), to industrial organisations before this time.
95. *Prikaz VSNKh SSSR*, no. 92, 21 October 1929.
96. See *Bolshaya Sovetskaya Entsiklopediya* (1st edition) Vol.43, col 773.
97. *Prikaz VSNKh SSSR*, no. 839, 15 February 1930.
98. *Prikaz VSNKh SSSR*, no. 897, 24 February 1930.

CHAPTER 5

1. G. Sakharov, N. Chernai and O. Kabakov, *Ocherki Organizatsii Tyazheloi Promyshlennosti SSSR* (Moscow, 1934) p. 66.
2. Ibid. These were the *glavki* for the basic chemical industry, the organic chemical industry, for gas and artificial liquid fuel and for the rubber industry.
3. Ibid.

4. *Predpriyatiya, Khozorgany i Uchrezhdeniya Narodnogo Kommissariata Tyazheloi Promyshlennosti* (Moscow–Leningrad, 1935).
5. *Za Ind.* 16 November 1930.
6. *Prikaz VSNKh SSSR*, no. 2356, 29 November 1930.
7. *Prikaz VSNKh SSSR*, no. 2358, 29 November 1930; he remained head until his political comeback in 1934, when he was succeeded by his deputy A. A. Armand – the first mention of the latter as head that has been found was in *Za Ind.* 26 November 1934.
8. *Za Ind.* 17 December 1930.
9. A statement of self-repudiation by him was published in *Pravda*, 20 November 1930.
10. In *Za Ind.* 16 November 1930 it was stated that NIS had been 'fruitful ground for the work of wreckers such as Ramzin'.
11. *Prikaz VSNKh SSSR*, no. 2376, 5 December 1930.
12. *Prikaz VSNKh SSSR*, no. 2422, 20 December 1930.
13. *Postanovlenie Prezidiuma Vysshego Soveta Narodnogo Khozyaistva SSSR*, no. 62, 9 February 1931, reprinted in *Direktivy i Formy po Sostavleniyu Tematicheskikh Planov i Kontrol'nykh Tsifr na 1932g. Nauchno-Issledovatel'skikh Uchrezhdenii VSNKh* (Moscow–Leningrad, 1931) pp. 14–15.
14. *Prikaz VSNKh SSSR*, no. 65, 7 February 1931.
15. *Prikaz VSNKh SSSR*, no. 362, 12 June 1931.
16. *Prikaz VSNKh SSSR*, no. 569, 23 August 1931.
17. See *SRIN*, no. 1 (1931) p. 232.
18. Twenty-four bodies controlled one institute alone; and twelve controlled two institutes.
19. *Prikaz NKTP SSSR*, no. 528, 29 July 1932.
20. Data from Appendix 1, Table A1.2, deflated by the price index from Appendix 2, Table A2.1.
21. See Appendix 1, Table A1.2; for the background to the switch in finance, see Chapter 7.
22. It is possible that the beginning of a review within the industrial research network may have been followed by its widening to include all the Soviet Union's research establishments.
23. *Za Ind.* 22 December 1932.
24. *Prikaz NKTP SSSR*, no. 976, 31 December 1932.
25. *Sotsialisticheskoe Stroitel'stvo SSSR* (Moscow, 1934) p. 420.
26. *SRIN*, no. 2–3 (1931) pp. 282–8.
27. *Prikaz NKTP SSSR*, no. 389, 23 April 1933.
28. The other two institutes in existence in 1931 – for glass and ceramics – were probably now under NKLegProm.
29. *Vtoraya Vsesoyuznaya Konferentsiya po Planirovaniyu Nauchno-Issledovatel'skoi Raboty v Tyazheloi Promyshlennosti. Rezolyutsii* (Moscow, 1933) p. 7.
30. A. A. Armand, 'Organizatsiya Nauchno-Issledovatel'skoi Raboty', *FNIT*, no. 1 (1933) p. 73.
31. *Postanovlenie Kollegii NKTP SSSR*, no. 87, 8 February 1933, reprinted in *SRIN*, no. 2 (1933) pp. 228–9; the resolution was put into effect in April when eighty-two institutes were linked to various plants, *Prikaz NKTP SSSR*, no. 418, 29 April 1933.

32. *SRIN*, no. 7 (1933) pp. 3-11.
33. *Prikaz VSNKh SSSR*, no. 569, 23 August 1931.
34. *SRIN*, no. 2 (1934) pp. 170-6.
35. *Predpriyatiya, Khozorgany...*; as previously stated, this source may not list all research institutes under NKTP at this time since some branches of industry are not included.
36. *SRIN*, no. 5 (1936) p. 166.
37. *SRIN*, no. 8 (1936) pp. 134-44.
38. Ibid., p. 143.
39. See the reports in *SRIN*, no. 5 (1936) pp. 166-7, *SRIN*, no. 8 (1936) pp. 152-5, *SRIN*, no. 9 (1936) pp. 159-61, *SRIN*, no. 10 (1936) p. 113.
40. See Appendix 1, Table A1.2.
41. A. V. Venediktov, *Organizatsiya Gosudarstvennoi Promyshlennosti v SSSR*, Vol. II (Leningrad, 1961) p. 581; for the transfer of the Central Research Institute for the Food and Agricultural Industry from NIS PTEU VSNKh to Narkomtorg, see *Prikaz VSNKh SSSR*, no. 1234, 17 April 1930.
42. Venediktov, *Organizatsiya...*, p. 581.
43. Decree of Sovnarkom of 11 September 1934, *SZ SSSR*, 1934, article 374.
44. *Trud v SSSR* (Moscow, 1936) p. 49.
45. For the reorganisation of NKLegProm, see the decree of TsIK and Sovnarkom of 17 July 1934, reprinted in *Za. Ind.* 18 July 1934; NIS NKSnab is named as one of the bodies which was to be sent for information a copy of the VSNKh directives to its research establishments for drawing up their 1932 plans, *Direktivy i Formy.. na 1932g..*, p. 2; there is no mention of such a sector in NKLesProm in the TsIK-Sovnarkom decree of 19 September 1934 on its reorganisation (*SZ SSSR*, 1934, article 371) and no such sector was included in NKPP's central administration on its foundation, decree of Sovnarkom of 11 September 1934 (*SZ SSSR*, 1934, article 374).
46. A decree of Sovnarkom RSFSR of 29 October 1931 (*SU RSFSR*, 1931, article 479) lists six institutes under VSNKh RSFSR; a list of the research establishments of VSNKh SSSR for November 1931 (*SRIN*, no. 2-3 [1931] pp. 282-8) states in addition that there were five institutes under VSNKh of the Ukrainian republic and single institutes under the VSNKhy of Belorussia and Uzbekistan.
47. Six Ukrainian institutes were transferred to NKTP in December 1932, *Prikaz NKTP SSSR*, no. 967, 29 December 1932, and four institutes formerly under VSNKh RSFSR were under NKTP in April 1933, *Prikaz NKTP SSSR*, no. 389, 23 April 1933.
48. See the decrees of Sovnarkom RSFSR of 29 October 1931 (*SU RSFSR*, 1931, article 479) and of 9 October 1934 (*SU RSFSR*, 1934, article 236).
49. *Chislennost' i Zarabotnaya Plata Rabochikh i Sluzhashchikh v SSSR* (Moscow, 1936) pp. 236-45.
50. *Industriya* 12 November 1937.
51. Statute approved by Sovnarkom on 10 November 1937, reprinted in *Finansovoe i Khozyaistvennoe Zakonodatel'stvo*, no. 1-2 (20 January

1938) pp. 15–17; see also the editorial in *Industriya* 17 September 1937.
52. See, for example, the statute of NKPP of 5 March 1938, reprinted in *Finansovoe i Khozyaistvennoe Zakonodatel'stvo*, no. 16 (10 June 1938) pp. 13–16, and of NKLegProm of 21 July 1938, reprinted in *Finansovoe i Khozyaistvennoe Zakonodatel'stvo*, no. 22–23 (20 August 1938) pp. 26–31.
53. K. Novikov, *Industriya* 26 April 1938, B. Volov, *Industriya* 11 May 1939.
54. *Kul'turnoe Stroitel'stvo SSSR* (Moscow, 1940) p. 238, B. Volov, *Industriya* 11 May 1939.
55. See Sovnarkom resolution no. 234 of 26 February on the structure and staffing of the new commissariat for the fuel industry, *Finansovoe i Khozyaistvennoe Zakonodatel'stvo*, no. 12 (30 April 1939) pp. 25–6.
56. See B. Lyubich, *Industriya* 29 June 1939, G. Sosunov, *Industriya* 17 February 1940.

CHAPTER 6

1. As we have seen in Chapter 5 NIS PTEU VSNKh became NIS VSNKh on the breaking up of PTEU in November 1930. The newly organised NKTP had a similar body. In December 1932 as part of a general reorganisation of NKTP's sectors, it was renamed TsNIS, the Central Research Sector (*Prikaz NKTP SSSR*, no. 886, 8 December 1932); by the beginning of 1934 the designation 'central' had been dropped. At the end of 1934 NIS was amalgamated with NKTP's Technical Propaganda Sector and became NIS-Tekhprop; finally in 1936, when it assumed responsibility for inventions, it became NISIZ. For the purposes of simplification the term NIS is used to refer to the sector throughout the years 1930–7.
2. It was reprinted in the first issue of the sector's bulletin, *Byulleten'* [Nauchno-Issledovatel'skogo Sektora PTEU VSNKh SSSR] no. 1 (1930) pp. 60–2.
3. The State Committee for the Coordination of Scientific Research (set up in 1961) and its successor the State Committee for Science and Technology (set up in 1965); on these bodies see the OECD publication, *Science Policy in the USSR* (Paris, 1969) pp. 56–61.
4. *Deyatel'nost' Nauchnoi Komissii Komiteta po Khimizatsii Narodnogo Khozyaistva SSSR pri SNK SSSR* (Leningrad, 1930) pp. 4, 8.
5. See *Izvestiya Nauchnoi Komissii Komiteta po Khimizatsii Narodnogo Khozyaistva SSSR pri SNK SSSR*, no. 1 (1931) pp. 13–21, 44–5, 48–50, *Organizatsiya Sovetskoi Nauki v 1926–1932gg. Sbornik Dokumentov* (Leningrad, 1974) pp. 120–1, *Vtoroi Plenum Komiteta po Khimizatsii Narodnogo Khozyaistva SSSR pri SNK SSSR. 28 Maya – 2 Iyunya 1930 Goda* (Leningrad, 1932) p. 313.
6. *SZ SSSR*, 1930, article 437, 5 August 1930.
7. See the reference by Bakh in *Izvestiya Nauchnoi Komissii Komiteta po Khimizatsii Narodnogo Khozyaistva SSSR pri SNK SSSR*, no. 1 (1931) p. 25.

8. Decree of Sovnarkom of 13 February 1931, *SZ SSSR*, 1931, article 113.
9. *Za Ind.* 16 November 1930.
10. See the case of the Leningrad Institute of Metals, ibid.
11. *Prikaz VSNKh SSSR*, no. 1508, 9 June 1930.
12. *Za Ind.* 24 January 1931.
13. *Postanovlenie Prezidiuma VSNKh SSSR*, no. 62, 9 February 1931, reprinted in *Direktivy i Formy... na 1932g...*, pp. 14–15.
14. A discussion of its planning role is deferred to Chapter 7.
15. *SRIN*, no. 2–3 (1931) pp. 282–8.
16. A. Ziskind, 'Kontrol'nye Tsifry Nauchno-Issledovatel'skikh Institutov Promyshlennosti na 1932g.', *SRIN*, no. 2–3 (1931) p. 280.
17. *Sotsialisticheskoe Stroitel'stvo SSSR*, (Moscow, 1934) p. 420.
18. *Prikaz VSNKh SSSR*, no. 420, 29 June 1931.
19. *Prikaz NKTP SSSR*, no. 528, 29 July 1932.
20. *Kommunisticheskaya Partiya Sovetskogo Soyuza v Rezolyutsiyakh i Resheniyakh S"ezdov, Konferentsii i Plenumov TsK*, Vol. III (Moscow, 1954) p. 183.
21. *Postanovlenie Kollegii NKTP SSSR*, no. 87, 8 February 1933, reprinted in *SRIN*, no. 2 (1933) pp. 288–9.
22. The exact date of the amalgamation is unknown, but the first reference to the new NIS-Tekhprop is in *Za Industrializatsiyu* in November.
23. *Prikaz VSNKh SSSR*, no. 324, 29 May 1931.
24. See, example, *Vtoraya Vsesoyuznaya Konferentsiya po Planirovaniyu Nauchno-Issledovatel'skoi Raboty v Tyazheloi Promyshlennosti. Rezolyutsii* (Moscow, 1933) p. 8.
25. *Prikaz NKTP SSSR*, no. 1632, 4 October 1936, reprinted in *FNIT*, no. 10 (1936) pp. 107–8.
26. For a detailed statement of its functions, see an addendum to *Prikaz NKTP SSSR*, no. 36, 13 January 1937, reprinted in *Finansovoe i Khozyaistvennoe Zakondatel'stvo*, no. 3 (30 January 1937) pp. 19–22.
27. *SRIN*, no. 9 (1936) p. 158.
28. See Loren R. Graham, *The Soviet Academy of Sciences and the Communist Party 1927–1932* (Princeton, 1967).
29. Ibid., pp. 80–141.
30. *Ustav Akademii Nauk Soyuza Sovetskikh Sotsialisticheskikh Respublik Utverzhden 23 Maya 1930 Goda* (Leningrad, 1930).
31. *Otchet o Deyatel'nosti Akademii Nauk Soyuza Sovetskikh Sotsialisticheskikh Respublik v 1933 Godu* (Leningrad, 1934) p. 60.
32. *SRIN*, no. 4 (1934) p. 164.
33. *Otchet.. Akademii Nauk... v 1933 Godu*, pp. 60–1.
34. See, for example, the functions of the commission on the durability of machinery, V. N. Treier, 'Problema Dolgovechnosti Mashin', *SRIN*, no. 7 (1934) p. 23.
35. *Otchet... Akademii Nauk... v 1933 Godu*, p. 11.
36. Ibid., p. 60.
37. *FNIT*, no. 5–6 (1934); the transfer of the Academy's headquarters and many of its institutes to Moscow was set in motion by a decree of Sovnarkom of 25 April 1934 (*SZ SSSR*, 1934, article 175).
38. *FNIT*, no. 5–6 (1934) p. 138.
39. *SRIN*, no. 2 (1935) p. 177.

40. Ibid., I. P. Bardin (ed.), *Ocherki po Istorii Akademii Nauk. Tekhnicheskie Nauki* (Moscow, 1945) p. 15.
41. G. M. Krzhizhanovskii, 'Akademiya Nauk v 1936g.', *FNIT*, no. 4 (1936) p. 8.
42. John Erickson, 'Radio-Location and the Air Defence Problem: The Design and Development of Soviet Radar 1934–1940', *Science Studies*, II (1972) pp. 248–50.
43. Krzhizhanovskii, *FNIT*, no. 4 (1936) p. 9.
44. On the need for the Academy to take over the general supervision of research in engineering, see E. A. Chudakov, 'Problemy Nauchno-Issledovatel'skoi Raboty v Oblasti Mashinostroeniya', *Sovetskaya Nauka*, no. 4 (1939) p. 72.
45. See *Sovetskaya Nauka*, no. 1 (1938) pp. 150–3, B. A. Keller, *Sovetskaya Nauka*, no. 4 (1938) pp. 143–4, *Plan Nauchno-Issledovatel'skikh Rabot Akademii Nauk Soyuza SSR po Osnovnym Problemam na 1941 God* (Moscow-Leningrad, 1940) p. 3.
46. V. S. Emel'yanov, 'U Istokov Atomnoi Promyshlennosti', *Voprosy Istorii*, no. 5 (1975) pp. 133–5, Arnold Kramish, *Atomic Energy in the Soviet Union* (Stanford, 1960) pp. 23–30.
47. B. V. Levshin, *Akademiya Nauk SSSR v Gody Velikoi Otechestvennoi Voiny* (Moscow, 1966).
48. This impression has been confirmed in conversations with Soviet scholars.

CHAPTER 7

1. (London, 1939).
2. N. I. Bukharin, 'Nauka i SSSR', *NR*, no. 11 (1927) p. 10.
3. A. V. Kol'tsov, *Lenin i Stanovlenie Akademii Nauk Kak Tsentra Sovetskoi Nauki* (Leningrad, 1969) pp. 254–6.
4. For the chemical industry, see *Tekhniko-Ekonomicheskii Vestnik*, no. 1 (1926) pp. 72–5; for organisations researching into minerals, *Tekhniko-Ekonomicheskii Vestnik*, no. 4 (1926) pp. 287–91 and no. 5 (1926) pp. 357–60.
5. Xenia Joukoff Eudin, Helen Dwight and Harold H. Fisher (eds), *The Life of a Chemist: Memoirs of Vladimir N. Ipatieff* (Stanford, 1946) p. 390.
6. P. S. Osadchii, 'Pervyi S"ezd Prezidiumov Gosudarstvennykh Planovykh Komissii', *NR*, no. 4 (1926) pp. 5, 10.
7. Cited by K. Troyanovskii, 'Ot Khozyaistvennogo Plana k Nauchnomu Planu', *Planovoe Khozyaistvo*, no. 4 (1930) p. 78.
8. *Protokol VSNKh SSSR*, 1926–7, article 319, 24 March 1927.
9. This implemented a resolution of STO and Sovnarkom of January 1927, *Ob"yasnitel'naya Zapiska k Proektu Edinogo Gosudarstvennogo Byudzheta Soyuza Sovetskikh Sotsialisticheskikh Respublik na 1927–1928 Byudzhetnyi God* (Moscow, 1928) p. 233.
10. *Prikaz VSNKh SSSR*, no. 53, 17 October 1927.
11. Troyanovskii, *Planovoe Khozyaistvo*, no. 4 (1930) p. 78.

12. A. Vangengeim and K. Troyanovskii, 'Organizatsiya Nauki v SSSR', *NR*, no. 11 (1929) pp. 27-32.
13. Troyanovskii, *Planovoe Khozyaistvo*, no. 4 (1930) pp. 87, 92.
14. S. I. Kaplun, 'O Planirovanii Nauchno-Issledovatel'skoi Raboty na Odnom iz Bol'nykh Uchastkov', *Planovoe Khozyaistvo*, no. 3 (1930) pp. 81-3.
15. Ibid., p. 101.
16. Troyanovskii, *Planovoe Khozyaistvo*, no. 4 (1930) p. 85.
17. See, for example, Kaplun, *Planovoe Khozyaistvo*, no. 3 (1930) p. 83.
18. See, for example, G. M. Krzhizhanovskii, 'Zadachi Sotsialisticheskogo Stroitel'stva i Nauchnye Rabotniki', *NR*, no. 5-6 (1928) pp. 3-13 and G. F. Grin'ko, 'Khozyaistvennoe Stroitel'stvo i Zadachi Nauchnykh Rabotnikov', *NR*, no. 12 (1928) pp. 3-12.
19. See L. M. Zak, 'Sozdanie i Deyatel'nost' VARNITSO v 1927-1932 Godakh', *Istoriya SSSR*, no. 6 (1959) pp. 94-107.
20. Each issue of *Izvestiya* which appeared at the time of the conference contained on average somewhat over a half a page of report and comment.
21. His report is discussed in detail in Loren R. Graham, 'Bukharin and the Planning of Science', *The Russian Review*, no. 1 (1964) pp. 135-48.
22. 8 April 1931.
23. Academician Ol'denburg, the long-serving secretary of the Academy of Sciences, was said to have spoken only of the need to collect data on the numbers of scientists and establishments and to improve the supply of foreign literature.
24. V. V. Kuibyshev, *Nauke - Sotsialisticheskii Plan* (Moscow-Leningrad, 1931).
25. N. I. Bukharin, 'Tekhnicheskaya Rekonstruktsiya i Tekushchie Problemy Nauchno-Issledovatel'skoi Raboty', *SRIN*, no. 1 (1933) p. 24.
26. For remarks by M. N. Pokrovskii presumably referring specifically to scientists working under Narkompros, see *Pyatnadtsatyi S"ezd VKP(b). Dekabr' 1927 Goda. Stenograficheskii Otchet*, Vol. II (Moscow, 1962) p. 1136.
27. *Organizatsiya Sovetskoi Nauki v 1926-1932gg.* (Leningrad, 1974) pp. 297-8.
28. V. D. Esakov, *Sovetskaya Nauka v Gody Pervoi Pyatiletki* (Moscow, 1971) p. 90.
29. Resolution of 7 August 1928, reprinted in *Resheniya Partii i Pravitel'stva po Khozyaistvennym Voprosam v Pyati Tomakh*, Vol. I (Moscow, 1967) pp. 750-5.
30. *Prikaz VSNKh SSSR*, no. 1063, 20 September 1928.
31. This was the period of the preparation of the 'December' variant, see E. H. Carr and R. W. Davies, *Foundations of a Planned Economy 1926-1929*, Vol. I (London, 1969) pp. 881-2.
32. M. Ya. Lapirov-Skoblo, 'Nauchno-Issledovatel'skaya Rabota v Promyshlennosti', *NR*, no. 1 (1929) p. 40.
33. P. S. Osadchii, 'Vedushchie Nachala Pyatiletnego Plana Narodnogo Khozyaistva SSSR', *NR*, no. 5-6 (1929) pp. 25-7.
34. The figures were presumably in 1927/28 rubles.

35. Lapirov-Skoblo, *NR*, no. 1 (1929) p. 40.
36. *Pyatiletnii Plan Nauchno-Eksperimental'noi Raboty v Svyazi s Rekonstruktsiei Promyshlennosti SSSR* nos. 1–; nos. 1–23 (Moscow, 1929) and no. 25 (Moscow, 1930) were consulted.
37. Two plans, including that for the All-Union Heat Engineering Institute headed by the ill-fated Ramzin, were well over 100 pages long and extremely detailed.
38. Percentage based on the data for the sixteen institutes which included such information in their plans.
39. M. Ya. Lapirov-Skoblo, 'Problema Nauchnykh Kadrov', *NR*, no. 11 (1929) p. 11.
40. *SZ SSSR*, 1929, article 268.
41. *Kommunisticheskaya Partiya Sovetskogo Soyuza v Rezolyutsiyakh i Resheniyakh S"ezdov, Konferentsii i Plenumov TsK*, Vol. II (Moscow, 1954) pp. 632–42.
42. See Eugène Zaleski, *Planning for Economic Growth in the Soviet Union 1918–1932* (Chapel Hill, 1971) pp. 117–23.
43. *Prikaz VSNKh SSSR*, no. 1710, 23 July 1930.
44. See M. Ya. Lapirov-Skhdlo, *Perspektivnye Plany Nauchno-Issledovatel'skikh Uchrezhdenii Promyshlennosti* (Moscow-Leningrad, 1931) p. 6, V. Milyutin, 'Vtoraya Pyatiletka i Zadachi Nauchnogo Fronta', *FNIT*, no. 4–5 (1932) p. 24.
45. *Pyatiletnii Plan Narodno-Khozyaistvennogo Stroitel'stvo SSSR*, Vol II (Moscow, 1930), Part I, pp. 161–2, 201–5, 311–12, 484–5, Part II, p. 19.
46. See Gosplan's guidelines for compiling the annual plan, for example – *Ukazaniya i Formy k Sostavleniyu Narodnokhozyaistvennogo Plana na 1937 God* (Moscow, 1936) pp. 417–21; see also P. Kagan, 'O Postroenii Ratsional'noi Klassifikatsii Nauchno-Issledovatel'skikh Rabot', *FNIT*, no. 5 (1936) p. 61.
47. The conference resolutions filled a volume 350 pages long, *Vtoraya Vsesoyuznaya Konferentsiya po Planirovaniyu Nauchno-Issledovatel'skoi Raboty v Tyazheloi Promyshlennosti. Rezolyutsii* (Moscow, 1933).
48. *Prikaz VSNKh SSSR*, no. 1063, 20 September 1928.
49. Lapirov-Skoblo, *NR*, no. 1 (1929) p. 40.
50. *Prikaz VSNKh SSSR*, no. 200, 13 November 1929.
51. See the guidelines for drawing up the plan which NIS issued, *Direktivy i Formy po Sostavleniyu Kontrol'nykh Tsifr na Nauchno-Issledovatel'skie Raboty Promyshlennosti na 1930–31g.* (Moscow, 1930).
52. Ibid., p. 3.
53. *Prikaz VSNKh SSSR*, no. 1352, 11 May 1930.
54. *Prikaz VSNKh SSSR*, no. 1697, 21 July 1930.
55. *Direktivy i Formy po Sostavleniyu Tematicheskikh Planov i Kontrol'nykh Tsifr na 1932g. Nauchno-Issledovatel'skikh Uchrezhdenii VSNKh* (Moscow-Leningrad, 1931).
56. *Prikaz VSNKh SSSR*, no. 440, 11 July 1931.
57. *Planirovanie i Operativnyi Uchet v Nauchno-Issledovatel'skikh Institutakh Promyshlennosti* (Moscow, 1932) p. 10.

58. A. Ziskind, 'Kontrol'nye Tsifry Nauchno-Issledovatel'skikh Institutov Promyshlennosti na 1932g.', *SRIN*, no. 2–3 (1931) pp. 277–81.
59. A. V. Ziskind, 'Kontrol'nye Tsifry Nauchno-Issledovatel'skikh Institutov Promyshlennosti na 1933g.', *SRIN*, no. 9–10 (1932) pp. 240–5.
60. The following description is based on the following sources: A. N. Bakh, *FNIT*, no. 1 (1934) pp. 71–3, *Chernaya Metallurgiya* 29 October 1940, J. G. Crowther, *Soviet Science* (London, 1936) pp. 22, 43–5, 87–8, 232 and N. Novikov, 'Planirovanie i Uchet Nauchno-Issledovatel'skoi Raboty', *FNIT*, no. 8 (1934)pp. 58–61.
61. On the associations see A. Klepikov, 'Nauchno-Issledovatel'skaya Khimicheskaya Assotsiatsiya TsNIS NKTP', *SRIN*, no. 8 (1933) p. 171, P. N. Lazarev, 'Rabota Svetotekhnicheskoi Sektsii Vsesoyuznoi Elektrotekhnicheskoi Assotsiatsii', *SRIN*, no. 8 (1934) p. 157 and *SRIN*, no. 4 (1933) pp. 173–5.
62. *Prikaz NKTP SSSR*, no. 1632, 4 October 1936, reprinted in *FNIT*, no. 10 (1936) pp. 107–8, *Industriya* 4 December 1938, and G. Sosunov, *Industriya* 17 February 1940.
63. See, for example, *Industriya*, 4 December 1938, Klepikov, *SRIN*, no. 8 (1933) p. 172 and I. Zalkind, 'Opyt Planirovaniya Nauchno-Issledovatel'skoi Raboty', *SRIN*, no. 5 (1935) p. 146.
64. Crowther, *Soviet Science*, p. 45.
65. *Chernaya Metallurgiya* 29 October 1940, D. Kravtsov, 'Vnedrenie Zakonchennykh Nauchno-Issledovatel'skikh Rabot v Promyshlennosti', *FNIT*, no. 2 (1935) p. 70, *SRIN*, no. 8 (1936) p. 140.
66. Esakov, *Sovetskaya Nauka* ...., p. 120.
67. Ziskind, *SRIN*, no. 9–10 (1932) p. 244.
68. See Stalin's speech to the Conference of Industrial Leaders in June 1931, Joseph Stalin, *Leninism*, Vol. II (London, 1933) pp. 440–2.
69. Edward Hallett Carr, *The Bolshevik Revolution 1917–1923*, Vol. II (London, 1952) pp. 303–5.
70. See, for example, Alexander Baykov, *The Development of the Soviet Economic System* (Cambridge, 1947) p. 169.
71. See the regulations for operating the system, *Osnovnye Polozheniya Khozrascheta v Nauchno-Issledovatel'skikh Uchrezhdeniyakh Promyshlennosti* (Moscow-Leningrad, 1931).
72. See a second set of regulations of January 1932, reprinted in *Planirovanie i Operativnyi Uchet* ..., pp. 57–64.
73. See, for example, D. A. Begak, *FNIT*, no. 9 (1933) p. 39, L. Reinberg, 'Pokonchit' s Otstavaniem Nauchno-Issledovatel'skikh Institutov Promyshlennosti', *FNIT*, no. 10 (1936) pp. 105–6, *SRIN*, no. 8 (1936) pp. 139–42, S. I. Vol'fkovich, *FNIT*, no. 9 (1933) p. 37.
74. *Prikaz NKTP SSSR*, no. 1632, 4 October 1936, reprinted in *FNIT*, no. 10 (1936) pp. 107–8.
75. B. Volov, *Industriya* 11 May 1939.
76. Ibid.
77. S. G. Strumilin, *K Metodologii Ucheta Nauchnogo Truda* (Leningrad, 1932) p. 5.
78. *Direktivy i Formy* ... *na 1930–31g.*, p. 4.

79. *Planirovanie i Operativnyi Uchet...*, pp. 10-11.
80. Ibid., pp. 35-6.
81. Strumilin, *K Metodologii...*, N. P. Suvorov, 'O Metodakh Izucheniya Effektivnosti Nauchnykh Rabot', *NR*, no. 12 (1928) pp. 23-33.
82. *Planirovanie i Operativnyi Uchet...*, p. 30.
83. Z. Grinberg and A. Rodin, 'Khozraschet v Nauchnykh Institutakh', *FNIT*, no. 4-5 (1932) p. 34.
84. See Graham, *The Russian Review*, no. 1 (1964) p. 144.
85. *Planirovanie i Operativnyi Uchet...*, pp. 29-30.
86. *Osnovnye Polozheniya Khozrascheta...*, p. 10.
87. Grinberg and Rodin, *FNIT*, no. 4-5 (1932) p. 34.
88. See Chapter 9 for information on bonuses to research workers for innovation.
89. See, for example, *FNIT*, no. 4-5 (1932) p. 52, B. Volov, *Industriya* 11 May 1939, and the NKTP decree no. 1632 of 4 October 1936 in *FNIT*, no. 10 (1936) pp. 107-8.
90. See, for example, *Za Ind.*, 16 November 1930.
91. N. I. Bukharin, 'Tekhnicheskaya Rekonstruktsiya i Tekushchie Problemy Nauchno-Issledovatel'skoi Raboty', *SRIN*, no. 1 (1933) p. 27.
92. This was stated to be a frequent occurrence by Grinberg and Rodin, *FNIT*, no. 4-5 (1932) p. 32.
93. See, for example, A. A. Armand, 'Organizatsiya Nauchno-Issledovatel'skoi Raboty: k Itogam II Vsesoyuznoi Konferentsii po Planirovaniyu Nauki v Tyazheloi Promyshlennosti', *FNIT*, no. 1 (1933) p. 74.
94. See, for example, A. A. Chernyshev, 'Ob Organizatsii Planirovaniya Nauchno-Issledovatel'skoi Raboty v Oblasti Elektrotekhniki', *SRIN*, no. 6 (1933) p. 124, *Industriya* 24 December 1938, B. Volov, *Industriya* 11 May 1939.
95. This was happening in the mid-1930s in the Ukrainian Academy of Sciences, V. P. Marchenko, *Planirovanie Nauchnoi Raboty v SSSR (na Opyte Ukrainskoi Akademii Nauk)* (Munich, 1953) pp. 10-11, and is suggested by Ruhemann's remarks on NKTP's Ukrainian Physical Technical Institute, M. Ruhemann, 'Soviet Physics in the 1930s', *New Scientist* 2 November 1967, p. 276.
96. *FNIT*, no. 1 (1934) p. 72.
97. E. Romanovskii, 'Nado Organizovat' Uchet Tematiki Nauchno-Issledovatel'skikh Rabot', *FNIT*, no. 3 (1934) p. 120.
98. See Appendix 1, Table A1.2.
99. A. Ziskind, 'Nauchno-Issledovatel'skie Kadry Promyshlennosti', *FNIT*, no. 7-8 (1932) p. 106.
100. Plan data from Ziskind, *SRIN*, no. 9-10 (1932) p. 241; actual from N. I. Bukharin, 'Nauchno-Tekhnicheskoe Obsluzhivanie Promyshlennosti', *SRIN*, no. 3 (1934) p. 6.
101. E. Kviring, *FNIT*, no. 5-6 (1934) p. 157.
102. See, for example, D. A. Gerasimov, 'Nauka i Ee Primenenie v Torfyanoi Promyshlennosti', *SRIN*, no. 2 (1934) pp. 84-5, Z. Grinberg and A. Rodin, 'Khozraschet v Nauchno-Issledovatel'skikh Uchrezhdeniyakh', *Kommunisticheskoe Prosveshchenie*, no. 2 (1932) pp. 10-17, K. Novikov, *Industriya* 26 April 1938 and *Industriya* 24 December 1938.

103. Grinberg and Rodin, *FNIT*, no. 4–5 (1932) p. 32.
104. Ibid., p. 31, *Planirovanie i Operativnyi Uchet*..., p. 37.
105. *Industriya* 24 December 1938.
106. Crowther, *Soviet Science*, p. 88.
107. Quoted by Chernyshev, *SRIN*, no. 6 (1933) p. 121.

CHAPTER 8

1. See, for example, A. P. M. Fleming and J. G. Pearce, *Research in Industry. The Basis of Economic Progress* (London, 1922), C. E. K. Mees, *The Organisation of Industrial Scientific Research* (New York, 1920).
2. Yu. N. Flakserman, 'Puti Stroitel'stva Nauchno-Issledovatel'skikh Institutov', *Tekhniko-Ekonomicheskii Vestnik*, no. 8 (1926) p. 526.
3. *Materialy Pervogo Vsesoyuznogo S"ezda po Voprosam Khimicheskoi Promyshlennosti* (Moscow, 1930) pp. 211–12.
4. The survey's findings and NTO's resolutions on it were reported in *Tekhniko-Ekonomicheskii Vestnik*, no. 10 (1926) pp. 682–5.
5. *Vsesoyuznaya Konferentsiya Predstavitelei Zavodskikh Laboratorii Metallopromyshlennosti. I-ya Moskva, 1928. Tezisy i Proekty Rezolyutsii k Dokladam* (Moscow, 1928) p. 13.
6. *TPG* 24–25 December 1927.
7. *Resheniya Partii i Pravitel'stva po Khozyaistvennym Voprosam v Pyati Tomakh*, Vol. I (Moscow, 1967) pp. 605–11.
8. *S"ezdy Sovetov Soyuza Sovetskikh Sotsialisticheskikh Respublik. Sbornik Dokumentov 1922–1936gg.* (Moscow, 1960) p. 121.
9. *Pyatnadtsatyi S"ezd VKP(b). Dekabr' 1927 Goda. Stenograficheskii Otchet*, Vol. II (Moscow, 1962) p. 1451.
10. *Vsesoyuznaya Konferentsiya*..., p. 13.
11. Ibid., p. 17.
12. Ibid., pp. 16–17.
13. *TPG* 10 August 1928.
14. *Resheniya Partii*..., pp. 753, 755.
15. *Prikaz VSNKh SSSR*, no. 893, 8 May 1929, reprinted in *TPG* 11 May 1929.
16. *Sotsialisticheskaya Rekonstruktsiya i Nauka. Sbornik Nauchno-Issledovatel'skogo Sektora PTEU VSNKh SSSR k XVI S"ezdu VKP(b)* (Moscow, 1930) p. 233.
17. Ibid.
18. *Prikaz VSNKh SSSR*, no. 200, 13 November 1929.
19. *Prikaz VSNKh SSSR*, no. 1495, 7 June 1930.
20. *Prikaz VSNKh SSSR*, no. 1909, 7 September 1930.
21. *Prikaz VSNKh SSSR*, no. 1508, 9 June 1930.
22. See, for example, V. Ya. Kurbatov's concluding remarks on his report to the Second Plenum of the Committee for Chemicalisation, *Vtoroi Plenum Komiteta po Khimizatsii Narodnogo Khozyaistva SSSR pri SNK SSSR. 28 Maya – 2 Iyunya 1930 Goda* (Leningrad, 1932) p. 308; also E. P. Frolov, *Osnovnye Zadachi Zavodskikh Laboratorii* (Moscow, 1933) p. 4.

23. *Prikaz VSNKh SSSR*, no. 2172, 19 October 1930.
24. *Postanovlenie Prezidiuma VSNKh SSSR*, no. 62, 9 February 1931, reprinted in *Direktivy i Formy po Sostavleniyu Tematicheskikh Planov i Kontrol'nykh Tsifr na 1932g. Nauchno-Issledovatel'skikh Uchrezhdenii VSNKh* (Moscow-Leningrad, 1931) pp. 14–15.
25. *Narodnoe Khozyaistvo SSSR* (Moscow-Leningrad, 1932) p. 546.
26. *ZL*, no. 8–9 (1932) p. 78.
27. Ibid., p. 76.
28. *SRIN*, no. 10 (1936) p. 110.
29. *ZL*, no. 8–9 (1932) pp. 76–7.
30. For two examples, see *Prikaz VSNKh SSSR*, no. 1024, 13 April 1930 and *Industrializatsiya Severo-Zapadnogo Raiona v Gody Vtoroi i Tret'ei Pyatiletok (1933–1941gg.)* (Leningrad, 1969) p. 145.
31. *Prikaz NKTP SSSR*, no. 617, 1 September 1932, reprinted in Frolov, *Osnovnye Zadachi*..., pp. 39–42.
32. See, for example, the laboratories at the AMO automobile plant, *ZL*, no. 10 (1932) pp. 15–16; at the 'Krasnoe Sormovo' engineering works, *ZL*, no. 1 (1932) p. 82, *ZL*, no. 9 (1933) p. 9 and *ZL*, no. 3 (1934) p. 271; at the Rostov agricultural engineering plant, *Nauchno-Tekhnicheskoe Obsluzhivanie Tyazheloi Promyshlennosti. Sbornik NISa i Tekhpropa NKTP k XVII S"ezdu VKP(b)* (Moscow-Leningrad, 1934) p. 261.
33. See, for example, N. I. Bukharin, 'Zavodskie Laboratorii i Puti Ikh Ukrepleniya', *ZL*, no. 1 (1933) pp. 5–9, the same author's, 'Fabrichno-Zavodskie Laboratorii – na Sluzhbu Osvoeniya Novoi Tekhniki', *ZL*, no. 7 (1933) pp. 3–4 and Frolov, *Osnovnye Zadachi*..., pp. 23–6.
34. *Prikaz NKTP SSSR*, no. 511, 26 July 1933, reprinted in *ZL*, no. 6 (1933) pp. 6–7.
35. *Kommunisticheskaya Partiya Sovetskogo Soyuza v Rezolyutsiyakh i Resheniyakh S"ezdov, Konferentsii i Plenumov TsK*, Vol. II (Moscow, 1954) p. 147.
36. Ibid., p. 212.
37. See, for example, *ZL*, no. 2 (1934) pp. 183–4, 188–9.
38. *Prikaz NKTP SSSR*, no. 435, 26 March 1934, reprinted in *ZL*, no. 5 (1934) p. 473.
39. *ZL*, no. 9 (1934) pp. 773–5.
40. *Prikaz NKTP SSSR*, no. 1082, 8 August 1934, reprinted in *ZL*, no. 8 (1934) pp. 775–6.
41. A. A. Armand (ed.), *Zavodskie Laboratorii Tyazheloi Promyshlennosti* (Moscow-Leningrad, 1935) pp. 8–9.
42. Ibid., and *Nauchno-Tekhnicheskoe Obsluzhivanie*....
43. See, for example, A. Klepikov, 'Moskovskaya Konferentsiya Zavodskikh Laboratorii Tyazheloi Promyshlennosti', *SRIN*, no. 7 (1936) p. 160; for criticism of their research activities see F. Donskoi, 'Po Zavodskim Laboratoriyam', *SRIN*, no. 3 (1935) pp. 60–78.
44. See, for example, the report of a December 1934 meeting of representatives of research institutes and factory laboratories, *ZL*, no. 1 (1935) pp. 113–15.
45. *Prikaz NKTP SSSR*, no. 873, 16 July 1935, reprinted in *Byulleten'*

Finansovogo i Khozyaistvennogo Zakondatel'stva, no. 22 (10 August 1935) pp. 11–12.
46. G. P. Efremtsev, *Istoriya Kolomenskogo Zavoda* (Moscow, 1973) p. 176.
47. *SRIN*, no. 5 (1936) p. 166.
48. See B. Shvyrev, *FNIT*, no. 3 (1935) p. 65, I. Abramov, 'Laboratorii Zavodov Chernoi Metallurgii', *SRIN*, no. 6 (1936) p. 150.
49. *SRIN*, no. 5 (1936) p. 166.
50. See Klepikov, *SRIN*, no. 7 (1936) p. 162.
51. *SRIN*, no. 8 (1936) p. 153.
52. *Za. Ind.* 6 July 1937.
53. I. Nikulinskii, *Industriya* 3 August 1940.
54. *Industriya* 26 April 1938.
55. *Industriya* 24 December 1938.
56. According to K. Novikov around 150 million rubles was to be spent on the laboratories of plants under NKTP in 1938, *Industriya* 26 April 1938.
57. See, for example, S. Burov, *Industriya* 3 January 1939, A. Lobovskii, *Industriya* 17 March 1940 and I. Nikulinskii, *Industriya*, 3 August 1940.
58. A. E. Fersman, '"Khoroshaya" Stat'ya', *SRIN*, no. 2–3 (1931) p. 180.
59. Donskoi, *SRIN*, no. 3 (1935) p. 70; but even here the construction of the laboratory had gone far from smoothly, *ZL*, no. 8–9 (1932) pp. 76–7.
60. See the following issues of *ZL*: no. 1 (1932) pp. 9–25, no. 5 (1933) p. 53, no. 5 (1934) pp. 470–1, no. 1 (1935) p. 114, no. 6 (1937) p. 661.
61. Klepikov, *SRIN*, no. 7 (1936) pp. 160–2.
62. *Vtoroi Plenum Komiteta po Khimizatsii*..., p. 306.
63. *Prikaz NKTP SSSR*, no. 55, 26 July 1933, reprinted in *ZL*, no. 6 (1933) pp. 6–7.
64. G. G. Povorin, 'Elementarnye Metodologicheskie Oshibki v Issledovatel'skoi Rabote', *ZL*, no. 1 (1935) p. 617.
65. E. Romanovskii, *FNIT*, no. 5 (1937) p. 124, *ZL*, no. 8 (1937) pp. 1033–4.
66. *ZL*, no. 11–12 (1932) p. 68, Bukharin, *ZL*, no. 7 (1933) p. 4, Klepikov, *SRIN*, no. 7 (1936) pp. 160–1, *ZL*, no. 6 (1935) p. 719, *ZL*, no. 8–9 (1938) pp. 901–4, I. Nikulinskii, *Industriya* 3 August 1940.
67. Frolov, *Osnovnye Zadachi*..., pp. 24, 52, Z. Slonimskii, 'Iz Opyta Khar'kovskogo Instituta Metallov po Instruktazhu Zavodskikh Laboratorii', *ZL*, no. 6 (1934) p. 572.
68. See, for example, A. Lobovskii, *Industriya* 17 March 1940.
69. Frolov, *Osnovnye Zadachi*..., p. 27; see also Klepikov, *SRIN*, no. 7 (1936) p. 163.
70. *Vtoroi Plenum Komiteta po Khimizatsii*..., p. 306.
71. Frolov, *Osnovnye Zadachi*..., p. 24; see also, I. Nikulinskii, *Industriya* 3 August 1940.
72. *ZL*, no. 1 (1935) p. 114.
73. Fleming and Pearce, *Research in Industry*..., pp. 77–9.
74. I. Nikulinskii, *Industriya* 3 August 1940.

CHAPTER 9

1. A. E. Fersman, 'Osnovnye Voprosy Organizatsii Nauchnoi Raboty', *Vestnik Akademii Nauk SSSR*, no. 10 (1936) p. 40.
2. *Nauchnye Dostizheniya v Promyshlennosti i Raboty Nauchno-Tekhnicheskogo Otdela VSNKh SSSR* (Moscow, 1925) p. 44.
3. *Chetyrnadtsataya Konferentsiya Rossiiskoi Kommunisticheskoi Partii (Bol'shevikov). Stenograficheskii Otchet* (Moscow-Leningrad, 1925) p. 174.
4. *Organizatsiya Nauki v Pervye Gody Sovetskoi Vlasti (1917–1925). Sbornik Dokumentov* (Leningrad, 1968) pp. 295, 334, 337.
5. *Protokol VSNKh SSSR*, 1926–7, article 319, 24 March 1927.
6. *TPG* 10 August 1928.
7. For a discussion of the term, see P. L. Kapitsa, *Teoriya, Eksperiment, Praktika* (Moscow, 1966) p. 7.
8. *Resheniya Partii i Pravitel'stva po Khozyaistvennym Voprosam v Pyati Tomakh*, Vol. I (Moscow, 1967) pp. 750–5.
9. *Prikaz VSNKh SSSR*, no. 1063, 20 September 1928, reprinted in *TPG* 21 September 1928.
10. *Pyatiletnii Plan Nauchno-Eksperimental'noi Raboty v Svyazi s Rekonstruktsiei Promyshlennosti SSSR*, no. 1, *Pyatiletnii Plan Rabot Vsesoyuznogo Teplotekhnicheskogo Instituta imeni Professora V. I. Grinevetskogo i K. V. Kirsha* (Moscow, 1929) pp. 16–17.
11. *Science Policy in the USSR* (Paris, 1969) p. 75.
12. *Za Ind.* 12 April 1931.
13. *Direktivy i Formy po Sostavleniyu Kontrol'nykh Tsifr na Nauchno-Issledovatel'skie Raboty Promyshlennosti na 1930–31g.* (Moscow, 1930) p. 4.
14. *Direktivy i Formy po Sostavleniyu Tematicheskikh Planov i Kontrol'nykh Tsifr na 1932g. Nauchno-Issledovatel'skikh Uchrezhdenii VSNKh* (Moscow-Leningrad, 1931) p. 5.
15. E. P. Frolov, 'Osnovnye Zadachi Zavodskikh Laboratorii', *ZL*, no. 8–9 (1932) p. 10.
16. Ibid.
17. *Planirovanie i Operativnyi Uchet v Nauchno-Issledovatel'skikh Institutakh Promyshlennosti* (Moscow, 1932) pp. 24, 53, *Vtoraya Vsesoyuznaya Konferentsiya po Planirovaniyu Nauchno-Issledovatel'skoi Raboty v Tyazheloi Promyshlennosti. Rezolyutsii* (Moscow, 1933); these short remarks are the only references to this study that have been found.
18. N. I. Bukharin, 'Tekhnicheskaya Rekonstruktsiya i Tekushchie Problemy Nauchno-Issledovatel'skoi Raboty', *SRIN*, no. 1 (1933) pp. 28–9.
19. *Vtoraya Vsesoyuznaya Konferentsiya...*, pp. 6–7.
20. *Postanovlenie Prezidiuma VSNKh SSSR*, no. 62, 9 February 1931, reprinted in *Direktivy i Formy... na 1932g....*, pp. 14–15.
21. *Postanovlenie Kollegii NKTP SSSR*, no. 87, 8 February 1933, reprinted in *SRIN*, no. 2 (1933) pp. 228–9.
22. *ZL*, no. 7 (1934) p. 666.
23. *Industrializatsiya Severo-Zapadnogo Raiona v Gody Vtoroi i Tret'ei Pyatiletok (1933–1941gg.)* (Leningrad, 1969) p. 151.

24. A. I. Aksenov, 'Perestroika Nauchno-Issledovatel'skoi Raboty Tsementnoi Promyshlennosti', *ZL*, no. 7 (1934) p. 663.
25. See, for example, the case of Ramzin's single-pass boiler, K. Ya. Bauman, 'Polozhenie i Zadachi Sovetskoi Nauki', *Vestnik Akademii Nauk SSSR*, no. 10 (1936) p. 23.
26. L. Reinberg, 'Pokonchit' s Otstavaniem Nauchno-Issledovatel'skikh Institutov Promyshlennosti', *FNIT*, no. 10 (1936) pp. 100–6, *SRIN*, no. 8 (1936) pp. 134–44.
27. *Prikaz NKTP SSSR*, no. 1632, 4 October 1936, reprinted in *FNIT*, no. 10 (1936) pp. 107–8.
28. See, for example, the bonuses for work on aluminium alloys paid to the staff of the All-Union Institute for Aviation Materials, *Prikaz NKTP SSSR*, no. 1584, 26 December 1934, reprinted in *Industrializatsiya SSSR 1933–1937gg. Dokumenty i Materialy* (Moscow, 1971) pp. 261–3.
29. *Industriya* 28 May 1938.
30. See, for example, M. Miz, *Industriya* 26 June 1940 and the issues of that newspaper for 26 May 1940 and 6 June 1940.
31. See the case of a process for producing liquid oxygen which had been developed by Kapitsa's institute, *Industriya* 10 September 1939, Kapitsa, *Teoriya...*, pp. 42–4, M. M. Levitin, *Industriya* 27 May 1940; see also B. Kil'chevskii and G. Mirkin, *Industriya* 21 August 1939.
32. *Industriya* 12 February 1939 and 26 May 1940, A. Kudashev, *Industriya* 17 July 1940.
33. I. V. Brenev (ed.), *Tsentral'naya Radiolaboratoriya v Leningrade* (Moscow, 1973) pp. 83–4.
34. For a western view on the importance of avoiding both an organisational and a spatial barrier, see the chapter by Jack Morton of the famous Bell Laboratories in David Allison (ed.), *The R&D Game: Technical Men, Technical Managers, and Research Productivity* (Cambridge, Mass., 1969) pp. 213–35.
35. E. P. Frolov, *Osnovnye Zadachi Zavodskikh Laboratorii* (Moscow, 1933) p. 17.
36. I. Shcherbakov, *Industriya* 9 May 1940.
37. Yu. N. Flakserman, *Promyshlennost' i Nauchno-Tekhnicheskie Instituty* (Moscow, 1925) p. 14.
38. *Poluzavodskaya Opytnaya Stantsiya Khimicheskogo Instituta im. L. Ya. Karpova* (Moscow, 1929).
39. *Pyatiletnii Plan...*, no. 13, *Pyatiletnii Plan Rabot Vsesoyuznogo Nauchno-Issledovatel'skogo Instituta po Udobreniyam imeni Professora Ya. M. Samoilova* (Moscow, 1929) p. 58.
40. I. Chuev, 'Promyshlennost' v 1928/29 Godu', *Vestnik Finansov*, no. 6 (1930) p. 116, *Ob"yasnitel'naya Zapiska k Proektu Edinogo Gosudarstvennogo Byudzheta Soyuza Sovetskikh Sotsialisticheskikh Respublik na 1927–1928 Byudzhetnyi God* (Moscow, 1928) p. 229.
41. M. Ya. Lapirov-Skoblo, 'Nauchno-Issledovatel'skaya Rabota v Promyshlennosti', *NR*, no. 1 (1929) p. 34.
42. *Pyatiletnii Plan....*, nos. 1–23, 25.
43. V. N. Kritsman, 'Osnovnye Problemy Razvitiya Mashiny v Sotsialisticheskom Khozyaistve', *SRIN*, no. 2 (1933) pp. 78–9.

44. A. A. Armand (ed.), *Nauchno-Issledovatel'skie Instituty Tyazheloi Promyshlennosti* (Moscow-Leningrad, 1935).
45. See, for example, A. N. Bakh and A. N. Frumkin, *Industriya* 26 October 1938, I. Konontsev, *Industriya* 26 April 1940.
46. For such facilities acting as technical servicing bodies for the rest of the institute, see *Materialy Pervogo Vsesoyuznogo S"ezda po Voprosam Khimicheskoi Promyshlennosti SSSR* (Moscow, 1930) p. 224, Armand (ed.), *Nauchno-Issledovatel'skie Instituty...*, p. 362.
47. Armand (ed.), *Nauchno-Issledovatel'skie Instituty...*, pp. 362, 887.
48. Ibid., pp. 173–4, *Materialy Pervogo Vsesoyuznogo...*, pp. 210, 224, *Tekhniko-Ekonomicheskii Vestnik*, no. 4 (1927) pp. 268–9.
49. *SRIN*, no. 8 (1936) p. 136.
50. For such an example from the chemical industry, see *Industriya* 23 November 1938.
51. *Prikaz NKTP SSSR*, no. 1632, 4 October 1936, reprinted in *FNIT*, no. 10 (1936) pp. 107–8.
52. See below.
53. On plastics see *Industriya* 8 May 1938, on machine-tools, Julian M. Cooper, *The Development of the Soviet Machine Tool Industry, 1917–1941* (unpublished PhD thesis: University of Birmingham, 1975) p. 303.
54. *Predpriyatiya, Khozorgany i Uchrezhdeniya Narodnogo Komissariata Tyazheloi Promyshlennosti* (Moscow, 1935).
55. Cooper, *The Development of the Machine Tool Industry*, p. 316.
56. *XVII Konferentsiya Vsesoyuznoi Kommunisticheskoi Partii (Bol'shevikov). Stenograficheskii Otchet* (Moscow, 1932) p. 77.
57. A. A. Armand (ed.), *Zavodskie Laboratorii Tyazheloi Promyshlennosti. Sbornik Soveta Zavodskikh Laboratorii NKTP SSSR* (Moscow-Leningrad, 1935).
58. See, for example, L. Lavrov, *Industriya* 9 February 1940, I. Konontsev, *Industriya* 26 May 1940.
59. Cooper, *The Development of the Machine Tool Industry*, p. 315.
60. *Science Policy in the USSR*, pp. 421–4.
61. E. A. Chudakov, 'Problemy Nauchno-Issledovatel'skoi Raboty v Oblasti Mashinostroeniya', *Sovetskaya Nauka*, no. 4 (1939) p. 74.
62. Cooper, *The Development of the Machine Tool Industry*, pp. 327–8.
63. See, for example, O. Yu. Shmidt, the vice-president of the Academy of Sciences, in *Otchet o Rabote Akademii Nauk SSSR za 1939g.* (Moscow-Leningrad, 1940) p. 9.
64. *NR*, no. 2 (1927) p. 126.
65. *Resheniya Partii...*, p. 751.
66. See *FNIT*, no. 6 (1933) p. 45, *ZL*, no. 1 (1935) p. 117.
67. *Industriya* 12 February 1939 and 27 August 1939.
68. For example, in the case of the process developed by Kapitsa's institute, which has already been referred to; see also A. N. Bakh and A. N. Frumkin, *Industriya* 26 October 1938, B. Volov, *Industriya* 11 May 1939, and *Industriya* for 12 February 1939 and 26 May 1940.
69. *Kommunisticheskaya Partiya Sovetskogo Soyuza v Rezolyutsiyakh i Resheniyakh S"ezdov, Konferentsii i Plenumov TsK*, Vol. III (Moscow, 1954) p. 432.

Notes to pp. 124-32   199

70. *Za. Ind.* 17 December 1930.
71. *Final Report of the Committee on Industry on Trade. Cmd. 3282* (London, 1929) pp. 214-18.
72. See *Science Policy in the USSR*, pp. 427-8.
73. For the importance of this criterion in the 1930s, see David Granick, *Management of the Industrial Firm in the USSR* (New York, 1954) pp. 150-7.
74. See, for example, Bakh's complaint in *FNIT*, no. 9 (1933) p. 43.
75. Armand (ed.) *Nauchno-Issledovatel'skie Instituty* ..., p. 533.
76. See, for example, R. W. Davies, 'Aspects of Soviet Investment Policy in the 1920s', in C. H. Feinstein (ed.), *Socialism, Capitalism and Economic Growth* (Cambridge, 1967) pp. 285-305.
77. Walter Arnold Rukeyser, *Working for the Soviets* (London, 1932) p. 266.
78. *Materialy Pervogo S"ezda* ..., p. 157.
79. Branch offices and bureaux were, in fact, established abroad, see *NR*, no. 4 (1927) p. 113, V. S. Lel'chuk, *Sozdanie Khimicheskoi Promyshlennosti SSSR: iz Istorii Sotsialisticheskoi Industrializatsii* (Moscow, 1964) p. 86.
80. *Postanovlenie Kollegii NKTP SSSR*, no. 87, 8 February 1933 reprinted in *SRIN*, no. 2 (1933) pp. 228-9.
81. For two examples from 1935, see *Predpriyatiya, Khozorgany* ..., pp. XX, XXII.
82. V. V. Kuibyshev, *Izbrannye Proizvedeniya* (Moscow, 1958) p. 218.
83. G. K. Orzhonikidze, *Industrial Development in 1931 and the Tasks for 1932. Report to XVII Conference of the Communist Party of the Soviet Union* (Moscow, 1932) p. 21.
84. *New Scientist* 2 November 1967, p. 277.
85. *Za Ind.* 17 December 1930.
86. *Za Ind.* 12 April 1931.
87. Reported in *SRIN*, no. 8 (1936) p. 142.
88. *Za Ind.* 16 November 1930.
89. *Postanovlenie Kollegii NKTP SSSR*, no. 87, 8 February 1933, reprinted in *SRIN*, no. 2 (1933) pp. 228-9.
90. Chudakov, *Sovetskaya Nauka*, no. 4 (1939) p. 73.
91. S. I. Vol'fkovich, *FNIT*, no. 9 (1933) pp. 36-7.
92. *Industriya* 23 November 1938.
93. See Alexander Vucinich, *Science in Russian Culture 1861-1917* (Stanford, 1970) p. 395.
94. For the view that the Soviet Union was almost totally dependent on imported technology in every sphere, see Anthony C. Sutton, *Western Technology and Soviet Economic Development 1917 to 1930* (Stanford, 1968) and *Western Technology and Soviet Economic Development 1930-1945* (Stanford, 1971).
95. R. A. Lewis, 'Innovation in the USSR; the Case of Synthetic Rubber', *Slavic Review* (forthcoming).
96. No comprehensive account of the industry in the interwar period has yet been written; material on the development of the industry can be found in *Istoriya Vtoroi Mirovoi Voiny*, Vols I & II (Moscow, 1973 and 1974), *The Soviet Aircraft Industry* (Chapel Hill, 1955), A. A. Velizhev,

Dostizheniya Sovetskoi Aviapromyshlennosti za Pyatnadtsat' Let (Moscow–Leningrad, 1932).
97. N. I. Bukharin, 'Tekhnicheskaya Rekonstruktsiya i Tekushchie Problemy Nauchno-Issledovatel'skoi Raboty', *SRIN*, no. 1 (1933) p. 28.
98. G. K. Ordzhonikidze, *Stat'i i Rechi*, Vol. II (Moscow, 1957) p. 517.
99. A. S. Yakovlev, *Tsel' Zhizni* (2nd edition) (Moscow, 1968) pp. 178–9, 273.
100. K. E. Bailes, 'Technology and Legitimacy: Soviet Aviation and Stalinism in the 1930s', *Technology and Culture*, XVII (1976) pp. 55–81.
101. Alexander Boyd, *The Soviet Air Force since 1918* (London, 1977) p. 41.
102. Asher Lee (ed.), *The Soviet Air and Rocket Forces* (London, 1959) pp. 30, 131.
103. Ibid., p. 131.
104. A. Sharagin (G. A. Ozerov), *Tupolevskaya Sharaga* (Frankfurt/M, 1971) pp. 11–29.
105. *Nauka i Tekhnika SSSR 1917–1927*, Vol. III (Moscow, 1927) p. 501.
106. V. B. Shavrov, *Istoriya Konstruktsii Samoletov v SSSR do 1938 Goda* (Moscow, 1969) p. 345.
107. *Materialy k Istorii TsAGI* (Moscow, 1968) pp. 43–4.
108. A. I. Shakhurin, 'Aviatsionnaya Promyshlennost' v Gody Velikogo Otechestvennoi Voiny (iz Vospominaniyi Narkoma)', *Voprosy Istorii*, no. 3 (1975) p. 136.
109. *Materialy k Istorii TsAGI*, pp. 7, 51.
110. Yakovlev, *Tsel' Zhizni*, p. 204.
111. *Aviatsiya i Kosmonavtika SSSR* (Moscow, 1968) p. 309, Yakovlev, *Tsel' Zhizni*, p. 158.
112. *Energeticheskaya, Atomnaya, Transportnaya i Aviatsionnaya Tekhnika, Kosmonavtika* (Moscow, 1969) p. 343; *The Soviet Aircraft Industry*, p. 24.
113. Shavrov, *Istoriya Konstruktsii Samoletov...*, pp. 452–3.
114. Ibid., pp. 321–5; at the end of 1927 Grigorovich's organisation was transferred to a Moscow factory, ibid., p. 379.
115. Ibid., pp. 377–89.
116. Ibid., p. 422.
117. M. Arlazarov, *Front Idet Cherez KB* (2nd edition, Moscow, 1975) p. 39.
118. Shavrov, *Istoriya Konstruktsii Samoletov...*, p. 421.
119. Ibid.
120. Yakovlev, *Tsel' Zhizni*, p. 82.
121. Shavrov, *Istoriya Konstruktsii Samoletov...*, p. 421.
122. Ibid., Yakovlev, *Tsel' Zhizni*, p. 82.
123. Shavrov, *Istoriya Konstruktsii Samoletov...*, pp. 421–2; according to Arlazarov, *Front Idet Cherez KB*, p. 40, it was broken down into only three independent organisations.
124. Shavrov, *Istoriya Konstruktsii Samoletov...*, p. 422.
125. *Istoriya Vtoroi Mirovoi Voiny*, Vol. II, p. 195.
126. Sharagin, *Tupolevskaya Sharaga*, pp. 24–9.
127. Nine are reported to have been formed in *Istoriya Vtoroi Mirovoi Voiny*, Vol. III (Moscow, 1974) p. 383.
128. Shavrov, *Istoriya Konstruktsii Samoletov...*, pp. 466, 526, 553, 571.

129. There are many references to such work, ibid., *passim*.
130. Ibid., pp. 396, 541-2.
131. Ibid., pp. 435, 498-9, 532.
132. For example, at the new Voronezh plant, ibid., p. 543.
133. Ibid., pp. 435, 499, 541-2, 561-2, V. Stepanchenko and V. Petrenko, *Kievskie Samoletostroiteli* (Kiev, 1970) pp. 52-5.
134. Yakovlev, *Tsel' Zhizni*, p. 84.
135. This account is basically drawn from A. I. Nekrasov, *FNIT*, no. 9 (1933) pp. 32-4.
136. Shavrov, *Istoriya Konstruktsii Samoletov* ..., p. 441.
137. *The Soviet Aircraft Industry*, p. 140.
138. Arlazarov, *Front Idet Cherez KB*, p. 46.
139. A. Magid, *Bol'shaya Zhizn'* (Moscow, 1968) p. 112.
140. For the example of Il'yushin and the starting up of production of the Il-2 at the Voronezh plant, see P. Y. Kozlov, *Ily Letyat na Front* (Moscow, 1976), p. 10; on the moving of Putilov's design team which designed the Stal'-2 passenger plane to factory no. 81, see Shavrov, *Istoriya Konstruktsii Samoletov* ..., pp. 453-4.
141. *Flying Review International*, XXII (1967) p. 450; for another example see Shavrov, *Istoriya Konstruktsii Samoletov* ..., p. 428.
142. Ibid., pp. 370-1, 422-5.
143. Magid, *Bol'shaya Zhizn'*, p. 144.
144. Shavrov, *Istoriya Konstruktsii Samoletov* ..., p. 500.
145. *Istoriya Vtoroi Mirovoi Voiny*, Vol. III, p. 383; for such close control elsewhere in the military industrial sector, see V. Emelyanov, 'O Vremeni, o Tovarishchakh, o Sebe. Zapiski Inzhenera', *Novyi Mir*, no. 2 (1967) pp. 88-90, 105-6.
146. Yakovlev, *Tsel' Zhizni*, pp. 90-100.
147. See the decree of NKTP SSSR, no. 1632 of 4 October 1936, reprinted in *FNIT*, no. 10 (1936) pp. 107-8.
148. *Materialy k Istorii TsAGI*, p. 50.
149. Kozlov, *Ily Letyat na Front*, pp. 8-9.
150. Yakovlev, *Tsel' Zhizni*, pp. 201-5; for a similar importance placed on R&D elsewhere in the commissariat for the defence industry in the late 1930s, see Emelyanov, *Novyi Mir*, no. 2 (1967) p. 61.
151. But not entirely; see Shavrov, *Istoriya Konstruktsii Samoletov* ..., pp. 484-5.
152. Arlazarov, *Front Idet Cherez KB*, pp. 92-3.
153. Shavrov, *Istoriya Konstruktsii Samoletov* ..., p. 422; subsequent comment by this author suggests revisions of this plan were made later, ibid., pp. 424-5.
154. Ibid., p. 469.
155. See, for example, ibid., pp. 390-1.
156. Yakovlev, *Tsel' Zhizni*, p. 77.
157. Ibid., pp. 65-89; for a general discussion of Osoaviakhim, see William E. Odom, *The Soviet Volunteers: Modernisation and Bureaucracy in a Public Mass Organisation* (Princeton, 1973).
158. *Tovarishch Komsomol. Dokumenty S"ezdov, Konferentsii i Plenumov TsK VLKSM 1918-1968*, Vol. I (Moscow, 1969) pp. 458, 464, *O Komsomole i Molodezhi* (Moscow, 1970) pp. 424-5.

202   Science and Industrialisation in the USSR

159. See the 1935 letter from the secretary of the Central Committee of the Komsomols A. V. Kosarev to the Central Committee of the Party and the Commissar for Defence, *Marsh Udarnykh Brigad. Molodezh v Gody Vosstanovleniya Narodnogo Khozyaistva i Sotsialisticheskogo Stroitel'stva 1921–1942gg. Sbornik Dokumentov* (Moscow, 1965) pp. 382–4.
160. Yakovlev, *Tsel' Zhizni*, pp. 82–100.
161. For Lavochkin's efforts to gain independence, see Arlazarov, *Front Idet Cherez KB*, pp. 53–60.
162. Shavrov, *Istoriya Konstruktsii Samoletov*..., pp. 424–5.
163. Ibid., p. 515.
164. See Magid, *Bol'shaya Zhizn'*, p. 127.
165. Shavrov, *Istoriya Konstruktsii Samoletov*..., p. 121.
166. Vaclav Nemecek, 'Polikarpov: the Prolific Pioneer', *Flying Review International*, XXIII (1968) p. 401.
167. Yakovlev, *Tsel' Zhizni*, p. 83.
168. A. Chesalov, 'General'nyi Konstruktor', *Aviatsiya i Kosmonavtika*, no. 9 (1963) p. 74.

CHAPTER 10

1. See Ronald Amann, Julian Cooper and R. W. Davies (eds), *The Technological Level of Soviet Industry* (New Haven and London, 1977).
2. For expenditure data, see Raymond Hutchings, *Soviet Science, Technology, Design. Interaction and Convergence* (London, 1976) p. 67.
3. See *Science Policy in the USSR* (Paris, 1969).
4. On this reform, see Loren R. Graham, 'Reorganisation of the Academy of Sciences of the USSR', in Peter H. Juviler and Henry W. Morton (eds), *Soviet Policy-Making* (London, 1967) pp. 133–59.
5. *Science Policy in the USSR*, pp. 52–61.
6. Ibid., pp. 231–40.
7. Ibid., pp. 426–7.
8. See Moshe Lewin, *Political Undercurrents in Soviet Economic Debates* (London, 1975).
9. Joseph S. Berliner, *The Innovation Decision in Soviet Industry* (Cambridge, Mass., and London, 1976) pp. 115–20, *Science Policy in the USSR*, pp. 465–9, 472–3.
10. *Science Policy in the USSR*, p. 469.
11. Ibid., pp. 458–64.
12. Berliner, *The Innovation Decision*..., pp. 130–47.
13. However, there has been debate about whether the new *ob"edineniya* should, like TsIAM, just produce the first batches of new designs or whether they should be fully involved in subsequent large-scale production, ibid., pp. 145–6.
14. A recent study, however, suggests that in some fields at least technical development has not been as successful as previously thought, Amann *et al.* (eds), *The Technological Level*..., pp. 45–7, 407–522.
15. *Science Policy in the USSR*, pp. 435–8.
16. *Final Report of the Committee on Industry and Trade, Cmd. 3282* (London, 1929) pp. 214–18.

APPENDIX 1

1. *The Measurement of Scientific and Technical Activities. Proposed Standard Practice for Surveys of Research and Experimental Development* (Paris, 1970).
2. *Otchet ob Ispolnenii Gosudarstvennogo Byudzheta za 1935 God* (Moscow, 1937) pp. 5–7, 90, 93, 124–5.
3. Ibid., pp. 90–3, 126–7.
4. For the details of this estimate, see R. A. Lewis, *Industrial Research and Development in the USSR 1924–1935* (unpublished PhD thesis, University of Birmingham, 1975) pp. 365–6.
5. See, for example, *Otchet ... za 1935 God*, pp. 72–89, 122–3; such expenditure can also be found in the reports of the 1931, 1932, 1933, 1934 and 1937 budgets.
6. See, for example, *Ob"yasnitel'naya Zapiska k Proektu Edinogo Gosudarstvennogo Byudzheta Soyuza Sovetskikh Sotsialisticheskikh Respublik na 1927–1928 Byudzhetnyi God* (Moscow, 1928) p. 235, *Otchet Narodnogo Komissariata Finansov Soyuza SSR ob Ispolnenii Edinogo Gosudarstvennogo Byudzheta Soyuza Sovetskikh Sotsialisticheskikh Respublik za 1929–30g.* (Leningrad, 1931) pp. 84, 86–9 and *Otchet ... za 1935 God*, pp. 58–67, 108–9.
7. See, for example, *Kul'turnoe Stroitel'stvo SSSR. 1935* (Moscow, 1936) pp. 258–61.
8. *SU RSFSR*, 1923, article 527, 8 July 1923; for subsequent Sovnarkom decrees see *SU RSFSR*, 1924, article 520, 4 April 1924 and *SZ SSSR*, 1929, article 349, 12 June 1929.
9. See *Ob"yasnitel'naya Zapiska ... na 1927–1928*, p. 233.

APPENDIX 2

1. For wage data for scientists, see Ya. Danilovich, 'K Voprosu ob Organizatsii Zarabotnoi Platy v Nauchno-Issledovatel'skikh Uchrezhdeniyakh', *FNIT*, no. 5 (1936) p. 67, Z. Grinberg and A. Rodin, 'Oplata Truda Nauchnykh Rabotnikov', *FNIT*, no. 10–11 (1931) p. 49, M. Ya. Lapirov-Skoblo, 'Nauchno-Issledovatel'skaya Rabota v Promyshlennosti', *NR*, no. 1 (1929) p. 38 and I. S. Samokhvalov, 'Chislennost'' i Sostav Nauchnykh Rabotnikov SSSR', *SRIN*, no. 1 (1934) p. 143.
2. For a full discussion of the available information, see R. A. Lewis, *Industrial Research and Development in the USSR 1924–1935* (unpublished PhD thesis, University of Birmingham, 1975) pp. 371–3.
3. There may, in fact, have been a fall in this period, see the price index for industrial production in A. Mendel'son (ed.), *Pokazateli' Kon"yunktury Narodnogo Khozyaistva SSSR za 1923/24–1928/29gg.* (Moscow, 1930) p. 96.

APPENDIX 3

1. The results were published in *Nauchnye Kadry i Nauchno-Issledovatel'skie Uchrezhdeniya SSSR* (Moscow, 1930).
2. See, for example, I. S. Taitslin, 'O Chislennom Sostave Sektsii Nauchnykh Rabotnikov', *NR*, no. 5–6 (1928) p. 85.

3. *Nauchnye Kadry*..., p. 12.
4. Ibid., pp. 13-14.
5. Elsewhere in the publication (pp. 17 & ff.) the more detailed information on speciality, nationality etc., is based on a total of 43,449 postholders; 18,213 of these were in research establishments.
6. Ibid., pp. 14-16.
7. Ibid., p. 30; this was also true in subsequent years, see I. S. Samokhvalov, 'Chislennost' i Sostav Nauchnykh Rabotnikov SSSR', *SRIN*, no. 1 (1934) p. 141 and *Kul'turnoe Stroitel'stvo SSSR v Tsifrakh; ot VI k VII S"ezdu Sovetov (1930-1934gg.)* (Moscow, 1935) pp. 152-3.
8. Samokhvalov, *SRIN*, no. 1 (1934) pp. 128-9, 141-3.
9. *Trud v SSSR* (Moscow, 1936) pp. 26-31.
10. For a detailed discussion of Soviet data on 'science and scientific services' see the OECD publication, *Science Policy in the USSR* (Paris, 1969) pp. 504-5.
11. *Nauchnye kadry*..., p. 72.
12. *Trud v SSSR* (Moscow, 1968) p. 24.

Index

Academy of Sciences:
  Imperial, 1–2, 4–5, 27, 37
  Soviet: 7, 9, 39, 63; statutes of, 74, 75, 77; Technical Council, 76–7; Technical Division, 27, 77–8; Technical Group, 27, 75–6; 1961 reform of, 146–7; Siberian Division of, 147; relations with Narkompros, 37–8; R&D at, 26–7, 34, 75–6, 147, 179n; and the industrial research network, 74-8, 188n
Aeroflot, 134
Agricultural engineering, 70, 108, 194n
Aircraft industry: organisation of, 134, 139; growth and technical level of, 132–3; R&D in, 133–8, 149, 177n; design and design organisations in, 24, 33, 125, 133–8; development facilities in, 34, 125, 136–9, 144; innovation in, 132–3, 136–42, 149
  factories of: no. 1 (formerly 'Dux'), 134, 136, 137; 'Krasnyi Letchik', 135–6; Menzhinskii (no. 39), 135, 137, 142; 'Tenth Anniversary of October', 137; Voronezh, 201n; no. 5, 137; no. 25, 134–5; no. 28, 135; no. 81, 201n
All-Union Committee of Standardisation, 53
All-Union Diesel Institute, 128
All-Union Electrical-Engineering Institute, 31, 77, 175n, 183n, 190n
All-Union Institute for Aviation Materials, 134, 197n
All-Union Society of Workers in Science and Technology for Assisting Socialist Construction (VARNITSO), 82
AMO (later Stalin) car plant, 52, 194n
Armand, A. A., 60–1, 184n
Asbestos mining, 129
Auto-Tractor Research Institute, 31

Bakh, A. N., 48, 81, 82, 96–7, 99
Balfour Committee, 128, 150
Baranov, P. I., 133
Bardin, I. P., 63
Bauman, K. Ya., 63, 78, 131
Belorussian SSR, 163n, 165n
Bernal, J. D., 79
Bonuses, 94, 121, 131, 197n
Brailo, G. P., 117, 118, 130–1
Budget, state: category 'science' of, 16–17, 151–2; expenditure on science from, 7, 9–10, 146, 151–7; expenditure on R&D in 1935, 15, 17; and funding of industrial research, 23, 46, 59–60, 92, 158–65
Bukharin, N. I.: on science and technology, 8–9; expulsion from the Politbureau, 53; posts in science, 51, 54, 57, 60, 72, 106; on the organisation of R&D, 30, 51–3, 69; on research planning, 79, 82, 94; on contract research, 96; on science at factories, 106–8; on development facilities, 126; on innovation, 118–19, 127; on the aircraft industry, 132–3, 136

Central Administration of State Industry (of VSNKh) (TsUGProm), 44
Central Bureau for the Realisation of Inventions, 48
Central Committee for Water Conservancy, 45
Central Geological Survey Institute, 34
Central Institute for Aero-Engine Construction (TsIAM), 134, 137, 149, 202n
Central Institute of Metals, 31, 35
Central Radio Laboratory, 77, 122, 148
Central Research Institute for Building Materials, 60
Central Research Institute for the Food and Agricultural Industry, 185n

205

206 Index

Central Scientific and Technical Council (of NTO VSNKh) 39, 42
Central Scientific and Technical Laboratory (of the War Department), 4, 20, 37
Central Scientific Research Council for Industry, 54, 67-8
Chelyabinsk, 109
Chemical industry: administration of, 56, 183n; R&D manpower in, 31-2; R&D in, 31, 33, 83, 177n; organisation of R&D under, 58, 61-2, 74; contract research in, 99; factory research in, 106-7, 112-13; development facilities in, 4, 123,,125-6, 198n; project organisations of, 24-5; and innovation, 116, 120, 132; and foreign technology, 33; in Germany, 1
Chemicalisation, 83; *see also* Committee for Chemicalisation
Chetverikov, I. V., 141
Chichibabin, A. E., 5
Chief Administration for the Northern Sea Route, 141
Chief Geological Survey Administration, 53
Chief Inspectorate of Weights and Measures, 53
Chubar', V. Ya., 9
Chudakov, E. A., 126, 131
Coal industry, 24, 70-1, 75, 108, 125
Committee for Chemicalisation, 48-9, 68, 74, 106
Committee for Higher Education, 175n
Committee for Invention Affairs, 45
Communist Party:
 Congresses: 8th, 6; 15th, 9-11, 22, 46, 83, 103; 16th, 130; 17th, 108
 Conferences: 14th, 115; 16th, 11, 50, 59; 17th, 30, 108, 130; 18th, 127
 Central Committee: 38; department of science of, 63, 78; plenums of, 86, 173n; resolutions of, 8, 53, 72, 103
 Central Control Commission: reviews of the R&D network, 11, 12, 22, 46-7, 120; and reform of industry, 53; on Second Five Year Plan, 72; and innovation, 118, 120
 Party Programme of 1919, 6
 Politbureau: 53, 139
Conference on Industrial Leaders, 191n
Congress of Soviets: 3rd, 8; 4th, 103; 5th, 11, 49
Construction and building materials, 60, 120

Consumer goods industries, 24, 32, 56, 64, 65
Contract research: introduction of, 80, 152; expansion of, 88, 91-2; at higher educational establishments, 174n; effects of, 95-6, 98-100, 148; and innovation, 120
Cooper, J. M., 126, 128
Council of Factory Laboratories, 106, 108
Council of Ministers, 68
Councils of Assistance, 43-4, 45, 115
Council of the National Economy of the Northern Oblast, 39
Crowther, J. G., 34, 90, 99

Defence industry, 26, 65, 70, 83, 110, 132, 149
Design: at institutes, 117, 133-6; at enterprises, 26, 109, 136; role of individuals in, 141-2; independent design organisations, 13, 24, 30, 133-8, 169-70
Development: finance of, 116, 144; organisation of, 117, 136-9, 144; attitude of industry towards, 127-8; role of the individual in, 141-2; plan for, 140-1
 facilities for: shortages of, 117-18, 122-8, 147; at institutes, 117, 123-5, 130-1, 136-7; at design organisations, 137; at factories, 26, 126, 128; independent development organisations, 24, 30, 125-6, 169-70; used for normal production, 34, 124-6, 147
Dnepropetrovsk, 174n
Dolgov, A. N., 42-4
Dzerzhinskii, F. E.; and scientists, 6; on science and technology, 8, 115; and NTO, 42-4, 51, 143-4; death of, 44, 182n

Electrical industry, 31, 32, 70, 107, 110, 122, 124
Electrification, 83; *see also* GOELRO
Elektrostal', 107
Engineering: and foreign technology, 31; R&D manpower in, 32; R&D at plants of, 26, 107, 109-10; coordination of R&D in, 188n; design in, 24, 26, 109; development facilities in, 124-6, 128; project organisations in, 24; innovation in, 131; *see also* Agricultural engineering; Electrical industry; Machine tool industry

Index 207

Fersman, A. E., 110, 114
First All-Union Conference for the Planning of Scientific Research, 90, 94, 99, 117, 130–1
First All-Union Conference of Representatives of Factory Laboratories of the Metal Industry, 103
First Five Year Plan, 8, 11, 24, 83, 130, 189n
First Gosplan Conference, 80
Flakserman, Yu. N., 44, 46, 101, 123
Food industry, 64
Foreign technology: import of, 2–3, 8, 11, 36, 48, 50, 125, 132, 145–6; Soviet R&D and, 31–3, 35, 128–30
Frascati Manual, 35, 151–2, 170
Frolov, E. P., 118, 123
Fuel industries, 32, 33, 70, 186n; see also Coal industry

Geological Committee, 1–2, 48, 53
Germany, 1, 4, 33
GEU (Chief Economic Administration of VSNKh), 42, 44
Gintsvetmet (State Institute of Non-Ferrous Metallurgy), 34, 131
Giprokhim (State Institute for Projecting Chemical Factories), 24
Gipromez (State Institute for Projecting New Metallurgical Factories), 24, 129
Glavkhim (Chief Administration of the Chemical Industry), 41, 52, 129
Glavkhimprom (Chief Administration for Inorganic Chemicals), 61
Glavki: 45, 47, 133; and the administration of industry, 40, 44, 49–50, 56, 59, 61–2, 65–6, 130, 146, 183n; and the control of R&D, 48–9, 59, 60–6, 73, 77, 109, 122, 130, 134; and the finance of research, 80, 84, 88; and the planning of research, 80, 92, 95, 103; and factory R&D, 103, 109, 112; and development, 116, 124–5, 139; and innovation, 116, 118–21, 123, 127, 140; and the import of foreign technology, 129, 146
Glavvtuz (Chief Administration for Higher Technical Educational Establishments), 73, 85
GOELRO (State Commission for the Electrification of Russia), 6, 48
Gorbunov, N. P., 37–40,
Gosplan (State Planning Commission), 52, 68, 84: and First Five Year Plan, 86, 87; and planning of science, 9, 80–2, 87, 98, 100; and the coordination of research, 79; and data on research manpower, 169–70
Grigorovich, D. P., 135, 137, 142
Groznyi Oil Trust, 62
GUAP (Chief Administration of the Aircraft Industry), 134, 140, 141
GUMP (Chief Administration of the Metals Industry), 65
GVF (Civil Air Fleet), 134, 141

Head (*golovnye*) institutes, 33, 57, 58
Higher educational establishments: employment in, 169–70, 172n; reorganisation of the pay system in, 15; funds for R&D in, 155n; R&D at, 3, 15-16, 62, 134, 136, 140, 175n; and five year plan for research, 83–4; see also Technical institutes; Universities

Ilyushin, S. V., 139, 201n
Industrial Party, 58
Institute for Chemical Physics, 33
Institute for Chemical Reagents, 34
Institute for Experimental Medicine, 2
Institute for the Study of the North, 183n
Institute of Telemechanics, 77
Ioffe, A. F., 7, 33, 57, 62, 65, 77, 90
Ipatieff, V. N., 41–3, 46, 51, 73, 74, 182n

Jasny, N., 17, 168

Kaftanov, S. V., 175n
Kaganovich, L. M., 53, 121
Kamenev, L. B., 51
Kapitsa, P. L., 197n
Karabash, 109
Karpov Chemical Institute, 7, 33, 52, 58, 62, 116, 123
KEPS (Committee for the Study of the Natural Productive Forces of Russia), 4–5, 39, 79–80
KGB (Committee of State Security), 135
Khar'kov, 39
Khar'kov Tractor Factory, 110
Khimstroi (Chemical Equipment Construction Company), 24
Khozraschet, 91, 94–9, 148
Kh. S. Ledentsov Society for Advancing the Experimental Sciences and Their Practical Application, 4
Kiev, 39
Kirov (pre-revolutionary Putilov) factory, 2, 26

208   Index

Kirovgrad, 109
Kocherigin, S. A., 135
Kola peninsula, 33
Kolomna railway engine works, 109
Komintern radio factory, 122
Komsomol, see VLKSM
Kosarev, A. V., 202n
'Krasnoe Sormovo' engineering works, 194n
Krasnouralsk, 109
Krzhizhanovskii, G. M., 11, 48, 76, 80
Kuibyshev, V. V.: 48, 108, 175n; head of VSNKh, 44; and administration of industry, 52; on R&D, 44–5, 52, 82, 130; and Ipatieff, 182n
Kviring, E. I., 98

Lapirov-Skoblo, M. Ya., 84–5, 88, 97, 123
Lavochkin, S. A., 142, 202n
Lebedev, P. N., 5
Ledentsov, Kh. S., 4
Lenin, V. I., 6, 37–8
Leningrad: 25, 26, 76, 107, 122, 135, 136; concentration of R&D in, 29–30, 123, 145; factory laboratories in, 102–3, 106; study of the plastics industry in, 120
Leningrad Electrophysics Institute, 77
Leningrad Institute of Metals, 187n
Leningrad Physical Technical Institute (formerly State Physical Technical Institute), 7, 18, 33, 52, 57, 65, 77, 96
Leningrad Plastics Institute, 120
Leningrad University, 174n
Likachev, I. A., 52
Lokomotivoproekt, 109
Luppol, I. K., 99

Machine-tool industry, 26, 31, 110, 125–6, 128
Magnitogorsk, 110
Malinovskii (deputy director of Gintsvetmet), 131
Martens, L. K., 46
Mekhanobr institute, 57, 58
Mendeleev, D. I., 5
Mezhlauk, V. I., 49, 52
Metallurgy: research manpower in, 31–2; R&D facilities in, 70–1; factory laboratories in, 102–3
   Ferrous: administration of, 55, 65; R&D in, 31; factory laboratories in, 107, 109–11, 113; project organisation for, 24
   Non-ferrous: reorganisation of R&D in, 108; factory laboratories in, 108–9, 112–13; development organisations in, 125
Mining Chemical Trust, 112–13
Ministry of Education (Tsarist), 2, 3, 7
Mnogotemnost', 96
Molotov, V. M., 108
Moorsteen, R., 168
Moscow: 76, 107, 134, 135, 136, 137, 138, 187n; concentration of R&D in, 29–30, 123, 145; factory laboratories in, 102
Moscow Aviation Institute, 134
Moscow Higher Technical College, 4, 7
Multiple-post-holding, 13, 169–70

Narkompros, (People's Commissariat of Education), 7, 29, 37–9, 83, 179n
Narkomtorg (People's Commissariat of Trade), 64, 185n
National Physical Laboratory, 1
Nekrasov, A. I., 76
New Economic Policy, 8, 40, 41
NIS (Scientific Research Sector): 60, 61, 77, 117, 118, 184; foundation of, 54; reorganisation of, 55, 186n; abolition of, 62–3, 65; and the administration of research, 56–9, 63, 77, 88–9, 185n; and the planning of research, 82, 86–7, 89–90, 97, 117–18; and the coordination of research, 67–74; and reorganisation of the R&D network, 60, 62; and the finance of research, 92; and factory research, 105–6; and development facilities, 117–18; and innovation, 119
NISIZ, 186n
NIS-Tekhprop, 186n
NKCherMet (People's Commissariat of Ferrous Metallurgy), 113
NKLegProm (People's Commissariat of Light Industry), 56, 64, 65, 184n, 185n
NKLesProm (People's Commissariat of the Timber Industry), 56, 64, 185n
NKMashinostroenie (People's Commissariat of Engineering), 25, 65
NKMestProm (People's Commissariat of Local Industry), 64
NKPP (People's Commissariat of the Food Industry), 64–5, 185n
NKPS (People's Commissariat of Communications), 39
NKRKI (People's Commissariat of Work-

ers' and Peasants' Inspection), 103, 116–18, 120
NKSnab (People's Commissariat of Supply), 64, 185n
NKTP (People's Commissariat of Heavy Industry): 56, 125; reorganisation of, 65–6; research expenditure of, 158, 163n; R&D network of, 21–5, 59–66, 71–3, 76, 124, 185n; and factory laboratories, 106–10, 195n; and research personnel, 111; and innovation, 119–21, 130–1
NKVD (People's Commissariat of Internal Affairs), 133
NKZdrav (People's Commissariat of Health), 39
NKZem (People's Commissariat of Agriculture), 39
North Caucasus Industrial Institute, 183n
North Caucasus Institute for Applied Chemistry, 99
Novikov, K., 110
Novosibirsk, 147
NTO (Scientific and Technical Department):
of VSNKh RSFSR and then VSNKh SSSR: 115, 143; foundation and early years, 38–40; role of, 73, 77; criticism of, 41–3; reform into NTU, 44–5; and the coordination of research, 79; and the finance of research, 39, 152; research network of, 39–40, 133; and factory laboratories, 101–2; and development facilities, 123
of the Ukrainian VSNKh, 39
of the Council of the National Economy of the Northern Oblast, 39
NTSy (Scientific and Technical Councils): 49, 50, 68; formation of, 42; and technical development, 46–7, 73; and the planning of research, 79–80, 84; and industry, 47–8, 116
Nuclear research, 78
NTU (Scientific and Technical Administration): 41, 56, 58, 67–8, 73–4, 77, 103; establishment of, 45; organisations under, 45, 48, 53–4, 183n; and technological development, 45–7, 52–3; role of, 47–9, 80; and the 1929 industrial reform, 50–3; abolition of, 53–4; and the planning of research, 80, 83–6, 88–9, 104; and factory laboratories, 104; and innovation, 115–16

Ob″edineniya: foundation of, 50–3, 119; division and reorganisation of, 55–6, 58–9, 119; abolition of, 61; and control of R&D, 51, 53, 57–61, 69–70, 89, 122, 130, 134; and the planning of research, 86, 89, 95; and factory laboratories, 105–6, 112–13; and development, 117–18; and innovation, 71, 119, 123; scientific-production, 122, 146, 148–9
OECD, 148
OGPU (Unified State Political Administration), 135
Ol'denburg, S. F., 189n
Ol'denburgskii, Prince, A. P., 2
Optico-mechanical industry, 58
Ordzhonikidze, G. K., 55, 108, 120–1, 130, 132–3
Osadchii, P. S., 84–5
Osoaviakhim (Society for the Promotion of Defence, Aviation and Chemistry), 140

Pavlov, I. P., 2, 4, 5
Petrograd, 4, 39
PEU (Planning and Economic Administration of VSNKh), 52–3
Physikalisch-Technische Reichanstalt, 1
Podzemgaz, 125
Pokrovskii, M. N., 10–11, 38, 49, 59
Powell, R. P., 168
Power Institute, 27
Polikarpov, N. N., 134–5, 137–8, 142
Productivity, 18–19, 92–4, 98–9, 111
Progressive Faction (of the State Duma), 2
Project organisations, 13, 24, 30, 116–17, 129–30
PTEU (Planning, Technical and Economic Administration of VSNKh), 53, 55, 67, 69
Purges, 19, 133
Putilov, A. I., 201n
Pyatakov, G. I., 41–3, 73, 143–4, 182n

Rabpros (Union of Workers in Education and Socialist Culture), 169
Radiolocation, 77, 110
Ramzin, L. K., 58, 184n, 190n, 197n
Research associations, 90, 96–7
Research Institute for Fertilisers, 34, 123, 131
Research Institute for the Musical Industry, 64

210 Index

Research Institute of Organic Semi-Products and Dyes, 124–5
Revolutionary Military Council of the USSR, 139
Richard, P. E., 135
Romanovskii, E., 97
Rostov agricultural engineering works, 194n
RSFSR (Russian Soviet Federative Socialist Republic), 30, 162–3n, 165n
Rudzutak, Ya. E., 48
Ruhemann, M., 130
Rukeyser, W. A., 129
Russian Association of Sciences, 37–8
Rykov, A. I., 8

Scientific Automobile Engineering Institute, 52
Scientific Chemical Pharmaceutical Institute, 116
Scientific societies, 3–4
Second All-Union Conference for the Planning of Research in Heavy Industry, 60, 72, 87, 118, 120, 132
Second Five Year Plan, 12, 16, 72, 87, 163n
Sector for Technical Propaganda, 72, 187
Sevastopol, 141
Sevtsvetmet, 108
Shavrov, V. B., 140
Shein, S. D. 43, 51
Siberia, 29
Society for a Moscow Scientific Institute, 4
Sindikaty, 49–50
Society of the Friends of the Air Force (ODVF) (later part of Osoaviakhim), 140
Solzhenitsyn, A. I., 19
Sovnarkom (Council of People's Commissars):
  of RSFSR, 7, 37–8
  of USSR: 8, 48, 68, 152, 175n; and control of science, 9; decree of August 1928 on research, 22, 29; 47, 83–8, 103, 116, 124; and research in higher education, 25; and R&D in the provinces, 29; and reorganisation of industrial research, 60; and academy of chemical sciences, 68; and Academy of Sciences, 9, 78; and planning of research, 83–8; and factory laboratories, 103–4; and innovation, 116; and development facilities, 124
Soyuztverdosplav, 125
Stalin, J. V., 8–9, 191n
Stalin (formerly AMO) car factory, 26
State Ceramic Institute, 34, 39
State Committee for the Coordination of Scientific Research, 74, 147, 186n
State Committee for Science and Technology, 68, 74, 147, 186n
State Electrical Trust, 183n
State Institute for Applied Chemicals, 34
State Optical Institute, 58, 120
State Physical Technical Institute (later Leningrad Physical Technical Institute), 52, 57
STO (Council of Labour and Defence), 53, 152
'Storming', 99
Strumilin, S. G., 93
Sukhoi, P. O., 135
Sul'kevich, I. E., 93–4
Supreme Expert Council, 47, 48, 50
Suvorov, N. P., 93
Sverdlov, V. M., 46, 51, 80, 102–4, 115, 127
Sverdlov, Ya. M., 182n
Sverdlovsk, 123
Svetlana works, 26, 107, 110
Synthetic rubber, 132

Technical Conferences, 43–5, 115
Technical Council (of NKTP), 65, 111
Technical institutes, 3, 15–16, 25, 63, 140; *see also* Higher educational establishments
Training of research personnel, 16, 18, 34, 151
Trotsky, L. D., 8–9, 45–6, 51, 182n
Troyanovskii, K., 81–2, 99
Trusts: 41, 49–50, 56; and control of research, 42, 59, 61–2, 89, 144; laboratories of, 88–9, 103–4, 112–13; and finance of research, 80; and development, 116; and innovation, 119, 123; and the import of foreign technology, 129
TsAGI (Central Aero- and Hydrodynamic Institute), 7, 18, 34, 76, 132–41
TsIK (Central Executive Committee of the USSR), 46
TsKB (Central Design Bureau), 135
TsNIIMash (Central Scientific Research Institute of Engineering), 124, 125

Index 211

TsNIS, 186n
Tsvetmetzoloto, 112
Tukhachevskii, M. N., 133
Tupolev, A. N., 134, 136–9
Turkmen, SSR, 163n, 165n

Ukraine, 10, 39, 103, 106, 163n, 185n
Ukraine Chemical Trust, 62
Ukrainian Physical Technical Institute, 33–4, 77, 99, 130, 175n
United Nations Industrial Development Organisation, 35
United States, 1, 18, 27–8, 32
Universities, 1, 3, 15–16, 174n; *see also* Higher educational establishments
Urals, 29, 106, 108
Urals Asbestos Trust, 129
Urals Institute for Non-Ferrous Metals, 109
Urals-Kuzbass combine, 29
Urals trust for non-ferrous metals, 62
Uzbek SSR, 163n, 165n

Vangengeim, A., 81, 99
Vavilov, N. I., 19
Vernadskii, V. I., 3, 5, 143
VLKSM (All-Union Lenin Communist Union of Youth) 140–1, 202n
Vnedrenie, 116, 118–20, 123
Volgin, V. P., 75–6
Voroshilov, K. E., 10, 139
Vostokstal', 62

Vsekhimprom, 56, 99
VSNKh (Supreme Council of the National Economy):
 of RSFSR, 7, 20, 37–41, 56, 64
 of USSR: central apparatus of, 44–5, 47, 50–7, 68, 75, 77; and administration of industry, 40–1, 44, 50, 59, 64; and industrial development, 45, 47, 50, 55, 73; and First Five Year Plan, 83, 87; and organisation of research, 29, 41, 54, 58–9, 69–70, 73; research network of, 8, 20–2, 46, 59, 71; funding of research under, 9, 59, 91–2, 97, 152, 158–61, 162n, 163n, 165n; and coordination of research, 67–71, 79; and planning of research, 29–30, 80, 83–90
 of the Ukraine, 39, 56, 64, 185n
 of other republics, 56, 64, 185n
Vucinich, A., 5

War Chemical Committee, 4, 20
War Department, 4, 37

Yakovlev, A. S., 136, 139–40, 141, 142
Yakovlev, Ya. A., 50
Yulin, A. I., 52, 129

Zaporozhstal', 110, 113
Zhukovskii, N. E., 4, 7, 20
Zhukovskii Military Air Academy, 134
Ziskind, A. V., 89–90, 93
Zolotarev, A. I., 84